Amazing Stories of the Space Age

AMAZING STORIES

OF THE SPACE AGE

TRUE TALES OF NAZIS IN ORBIT, SOLDIERS
ON THE MOON, ORPHANED MARTIAN ROBOTS,
AND OTHER FASCINATING ACCOUNTS FROM
THE ANNALS OF SPACEFLIGHT

ROD PYLE

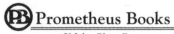 Prometheus Books

59 John Glenn Drive
Amherst, New York 14228

Cover images © NASA, PhotoDisc
Cover design by Nicole Sommer-Lecht
Cover design © Prometheus Books

Inquiries should be addressed to
Prometheus Books
59 John Glenn Drive
Amherst, New York 14228
VOICE: 716–691–0133 • FAX: 716–691–0137
WWW.PROMETHEUSBOOKS.COM

21 20 19 18 17   5 4 3 2 1

Library of Congress Cataloging-in-Publication Data

Names: Pyle, Rod.
Title: Amazing stories of the space age : true tales of Nazis in orbit, soldiers on the
    moon, orphaned martian robots, and other fascinating accounts from the annals
    of spaceflight / by Rod Pyle.
Description: Amherst, New York : Prometheus Books, 2016. | Includes index.
Identifiers: LCCN 2016032507 (print) | LCCN 2016041376 (ebook) |
    ISBN 9781633882218 (pbk.) | ISBN 9781633882225 (ebook)
Subjects: LCSH: Aeronautics—History | Astronautics—History. | Aeronautics and
    state. | Astronautics and state.
Classification: LCC TL670 .P95 2016 (print) | LCC TL670 (ebook) |
    DDC 629.409—dc23
LC record available at https://lccn.loc.gov/2016032507

Printed in the United States of America

CONTENTS

FOREWORD

I n my bedroom hangs a large oil painting that doesn't really belong there. It shows a shadowy, cylindrical spacecraft stealthily orbiting the Earth, about to dock with another mysterious space cylinder. The painting is from the early 1960s. I purchased it in an antiques store years ago, in a small town with major defense contractors nearby. The painting depicts what was once a highly secret US Air Force space project: the Manned Orbiting Laboratory or MOL. Very few illustrations of the MOL were ever made public. And the story behind the project, which has remained classified until recently, is fascinating. The air force once had secret plans to fly a space station of its own. Hardware was built and readied for launch, and astronauts were selected and trained for covert space missions. So what happened? You are about to read all about this and many other amazing stories of the space age in this wonderful book by Rod Pyle.

Amazing Stories of the Space Age is about the most mysterious and intriguing episodes of the history of space exploration—its under-cover projects, grandiose dreams, odd spinoffs, and muffled dramas. But rather than being tales of fiction or bogus conspiracy theories, the amazing stories presented here are all true, thoroughly researched, and expertly described. Rod Pyle has an encyclopedic knowledge of every-thing space—I love all of his books about Mars exploration—and his knowledge, enthusiasm, and humor shine through again in this book.

You are about to embark on a journey through time, all the way back to bygone eras of spaceflight, when spacecraft were still called rocketships and space stations would be shaped like wheels. The chapters convey the ominous angst of World War II and the Cold War, but also capture the boldness, optimism, and promise of the beginnings of spaceflight. You will read, with both wonder and nostalgia, accounts of the fantastical projects that were an integral, if not defining, part of the dawn of the space age. Each chapter is dedicated to a project, concept, or iconic product that, even if never fully realized, played an important role in shaping the history of space exploration and our relation to it. Pyle shows us that this history is not just the sum of what humans actually achieved in space, but also of what they never did or, maybe more accurately, haven't done yet.

While the amazing stories in this book have their specific context in history, many remain remarkably relevant, even timely. Take the moon base concepts of Project Horizon or the LUNEX Project, or Wernher von Braun's vision of inflatable space stations, or the advanced propulsion system of Project Orion. We should view these not so much as fields of failed dreams, but instead as the pioneering foundations of plausible developments about to be realized.

Chapter 3, which discusses von Braun's early plan to get humans to the red planet, is my favorite because it relates directly to my own lifelong dream of seeing humans explore Mars. During our recent Northwest Passage Drive Expedition in the Arctic, my teammates and I drove the HMP Okarian—a souped-up Humvee fitted with tracks, standing in as a future pressurized rover for Mars—across several hundred kilometers of frozen polar desert before finally reaching the barren, rocky shores of Devon Island, aka "Mars on Earth." As Pyle recounts in his book, von Braun had envisioned that the first humans on Mars would journey, in tracked vehicles, from one of Mars's snow-covered poles to its rocky equatorial regions. . . . In a way, our Arctic trek was déjà vu.

Illustrating Pyle's stories is an amazing collection of pictures and diagrams, some of them never published before. They are an integral part of what makes this book a real treasure.

Pascal Lee
Director, Mars Institute
Planetary Scientist, SETI Institute

NASA Ames Research Center
Moffett Field, California
September 2016

CHAPTER 1

NAZIS IN SPACE: PROJECT SILVERBIRD

CLASSIFIED: *TOP SECRET IN NAZI GERMANY, DECLASSIFIED AFTER THE END OF WORLD WAR II*

A *scenario:*

March 12, 1945, is a blustery day in Manhattan. Couples are strolling, enjoying the early spring weather. Earnest men dash across crowded boulevards; wool suits, ties, and fedoras are the uniform of the day. It is only 4:00 p.m., but the sidewalks are already shadowed canyons on Wall Street. The district is packed with those departing from work early, eager to begin the trek to the boroughs and home.

Most people are dashing to the subway, while others are engaged in animated conversation as they walk in pairs. The noises of urban life almost drown out the soft, twin pops that echo down the busy avenues, reverberating from the endless expanse of concrete and glass. A few look around, wondering what might have created the odd sound—it was too deep to be a backfiring taxi; it sounded almost like distant artillery. Nobody thinks for a moment that it might signal a few tons of explosive death falling into dense air high above the metropolis. Far downrange, a machine from the future glides silently onward, seeking escape from the impending cataclysm.

Then, a blinding flash of light heats the street to incandescence. Within seconds a shock wave devastates a two-mile-wide section of the city, shattering windows, gutting skyscrapers, and devastating multiple city blocks below. The Chrysler Building and Empire

State Building are rendered skeletal, windowless wrecks. Fires rage unchecked, and an estimated 300,000 die within moments. Twice that number are injured. Manhattan is a ruined inferno, its streets scattered with the smoldering forms of the dead and dying. And an invisible enemy—radiation—will stalk the city for weeks.

High above, in the tranquil blackness of space, a lone German pilot attempts to radio his success to Axis ships hiding in the Atlantic, having shaken off his awe at the utter destruction he has wrought. He is mildly surprised to find that his radio no longer functions, but that is of little consequence. His craft, the *Silbervogel Amerika Raketenbomber*, will cross the United States in record time at Mach 3.4. Soon he will land on Japanese-held territory in the Pacific, and will later be awarded the Reich's highest honor when he returns home by conventional aircraft. His silver spaceship, the bringer of war to American shores, will follow, lashed to the deck of a Japanese aircraft carrier. In Berlin, military planners are certain that World War II will soon be over, and the thousand-year Reich will emerge triumphant.

Of course, this nightmare never occurred outside the overheated minds of a handful of Nazi leaders, a small crew of aerospace engineers, and a brilliant rocket designer. But the Germans did work diligently for a time to develop a nuclear weapon, and Eugen Sanger and his partner Irene Bredt did develop detailed plans for a suborbital skip bomber ultimately intended to bomb Manhattan and other US targets, called the *Silverbird*.

Sanger was the driving force behind the rocket plane project. Born in 1905, he studied civil engineering as a young man until a stunning new book grabbed his imagination. Hermann Oberth, a German physicist and engineer, had written *Die Rakete zu den Planetenräumen* (*By Rocket into Planetary Space*) in 1923. Oberth was an early rocketry pioneer and saw the future clearly: rockets would one day allow people to travel into space. Many scoffed, as people will do when encountering new and visionary ideas, but Sanger, like his future rival Wernher von Braun, was fascinated by the book, and immediately altered his career trajectory to pursue aeronautics.[1]

Fig. 1.1. Aeronautical engineer Eugen Sanger and his future wife, mathematician Irene Bredt, work on the *Silbervogel* (*Silverbird*) rocket bomber early in WWII.

Sanger joined a group of brilliant young German engineers, the VfR or "Society for Space Travel," that had begun experimenting with rockets. Other nations had their own devotees of rocket propulsion—notably the US and USSR—but the German amateur rocketeers were well organized and made swift progress through the 1930s. As a group they followed the ideas of Oberth, and Sanger and von Braun became enthusiastic devotees. At a time when liquid-fueled rockets were a mere curiosity—Robert Goddard had pioneered the technology in the United States in the 1920s—it was an extremely dangerous endeavor, and experimentation frequently ended in disfiguring explosions. But from such determined origins spring great things, and humanity's reach into space was born in Goddard's workshop and the VfR's fiery experimentation.

In 1932, Sanger joined both the fledgling Nazi party and its elite paramilitary SS—membership in both organizations was beneficial to engineers and scientists seeking to advance their technical careers, especially if the projects in which they were interested had military applications. Like Oberth before him, he wrote of rocket-powered flight for his graduate thesis, which was rejected, as Oberth's had been, for being "too fanciful." Again following in Oberth's footsteps, he later published the treatise as a book.

Sanger's involvement with the Third Reich was preceded by a series of articles he had written for a German aerospace journal about rockets and flight in 1935–36. He was soon engaged to design ramjet engines for the Luftwaffe. But Sanger's passion for rockets could not be long suppressed.

Not far away, Wernher von Braun was designing the V-2 ballistic missile that would soon devastate targets in major European cities. His liquid-fueled rocket had a limited range—about 225 miles maximum—and could carry an explosive payload of only about 2,200 pounds, but was an effective terror weapon. By 1944, the V-2 began a campaign of aerial bombardment that, by war's end, launched 3,000 of the unstoppable weapons against Germany's enemies in Europe, primarily London and Antwerp.[2]

Sanger's project had a more daring goal, however, that attracted some within the German leadership to support his efforts. Since 1942, and with origins as far back as 1938, the Luftwaffe had been supporting a project to enable the bombing of the United States from Germany, or nearby conquered territories. The project was called the Amerika Bomber, and would take various forms as the country slid further into World War II.[3] Conventional long-range planes were proposed and prototyped; a "piggyback" ramjet bomber, carried aloft by a conventional plane, was designed. Sanger had larger ambitions, and worked on plans for a rocket-powered bomber he had long called *Silbervogel* (*Silverbird*) that would, in essence, bomb the United States from space. The *Silverbird* would never advance beyond planning and the testing of small models, but it was an inspired design for the 1930s/1940s.

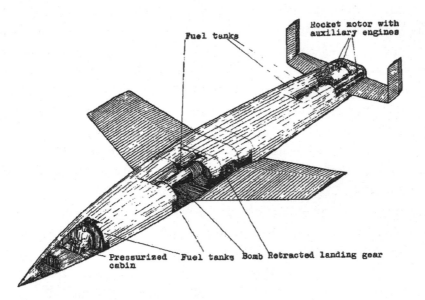

Fig. 1.2. Schematic view of the *Silverbird* in flight. The bomb bay is not shown, but the bulk of the rocket bomber is comprised of fuel tanks. This illustration is from a postwar US Army report. Image from DOD/US Army.

The rocket-powered bomber would be a ten-ton winged craft (100 tons when fueled), the development of which, even in the limited program—calculations, schematics, and wind tunnel tests—would pioneer ideas and technological developments that would later find their place in the future of spaceflight. For starters, the rocket engine, enclosed in the rear of the fuselage, was to be regeneratively cooled, using rocket fuel circulated through the rocket engine's nozzle to prevent it from being melted by the hot exhaust plume. This was a daring design in the day, causing many engineers to furrow their eyebrows at the thought of combustible liquids moving past red-hot metal surfaces, but it ultimately became the standard of most large rocket engines. The body of the rocket plane had wings, but also utilized what would become known as a "lifting body" principle, wherein the entire airframe generated additional lift. It was also flat-bottomed to allow it to skip along the upper atmosphere.[4]

Like the US Air Force's X-15 suborbital rocket plane, which did not fly until 1959, most of the *Silverbird*'s streamlined length contained tanks to hold fuel. It would begin flight by being pushed down a two-mile-long track by a "caboose," a larger, separate rocket motor that would accelerate the rocket plane to about 1,100 mph by the time it lifted from the track about ten seconds later. The "caboose" would stay on the ground, arrested at the end of the track and returned to its starting point for reuse.

The rocket engine in the rear of the *Silverbird* would burn kerosene and liquid oxygen to produce its own thrust, firing for four to eight minutes with an estimated thrust of an incredible (and doubtless optimistic) 220,000 pounds[5] (for comparison, the X-15's maximum thrust was about 57,000 pounds; the V-2 created 56,000 pounds). Sanger and Bredt calculated, again probably optimistically, that the rocket plane would attain speeds of somewhere between Mach 13 and Mach 20, depending on the weight of the payload and the mission profile (these were slide-rule estimates; had it been developed into a flight prototype, the numbers would likely have been quite different). The maximum altitude would have been between thirty and ninety miles, the latter of which was well above the generally accepted definition of space, which is about sixty-two miles. Once again, this was a region not experienced by human pilots until the advent of the X-15, which had the advantage of being hauled to a significant altitude by a B-52 carrier aircraft before igniting its own rocket engine. In the late 1930s, the altitude record for powered flight stood at about eleven miles.[6]

Sanger reasonably suggested that the pilot be placed in a reclining position, to better endure the 5–10 g of acceleration that would be experienced under power.

After the initial boost, the *Silverbird* would be carried by its momentum and increase its range by bouncing off the upper atmosphere, much like a properly tossed stone skipping on a lake. The flat wings and underbody, when appropriately angled, would skip off the denser air below. Each rebound would result in an initial altitude gain of about twelve miles, with each "skip" being somewhat shallower as the rocket plane lost energy. Eight such bounces were estimated before the rocket plane (with the fuel exhausted, it would actually be an unpowered glider) would settle into steady flight. It was considered unlikely that the craft would be able to complete a full trip around the globe to Germany, but since Japan had become a military ally in 1939, the *Silverbird* could land in a Japanese-held territory in the Pacific. It would then be transported back to Germany by ship, or possibly under its own power—Sanger suggested the construction of refueling depots throughout Axis territory, complete with launch facilities, to facilitate this. Depending on where it landed, the rocket plane would have flown a distance of nearly 9,000–10,000 miles during its primary mission. In scenarios where there was no safe landing site available, Sanger suggested a "sacrificial" bomb mission. He wrote, "Since the aircraft gains altitude rapidly after bomb release in a point attack, the pilot can, at the end of this brief climb, parachute from the plane and destroy the empty aircraft to keep it from getting into the hands of the enemy. He will land a few km. away from the point of the impact of his bombs, and be captured."[7] Easier to say if you are the designer and not the pilot.

The critical part of the flight profile would be the bomb run over the United States, and Manhattan was mentioned as a primary target. The rocket plane would time a dip into the atmosphere to coincide with its travel toward New York, and release its bomb payload, up to 8,000 pounds,[8] to glide to the target. While any attempt at accuracy at this altitude and speed was unlikely to succeed, if the explosive landed anywhere that was densely populated, it would mean a huge propaganda victory for the Reich. For very high altitude bombing, or when a target was not visible due to cloud cover, Sanger rather incredibly suggested that the pilot could navigate via the stars, clearly visible from the *Silverbird*'s intended altitudes, to release the bomb at the proper time and place.

Sanger continues, "Since the target, for the distances involved, will not be visible, the release on an area will be aimed indirectly, e.g. by celestial navigation. Thus it is independent of weather and visibility at the target. Because of this, it does not reach the accuracy of release on a point, and we must expect spreads of several kilometers. So with aerial bombing one cannot hit particular points, but rather a correspondingly large area, with sufficient probability. To achieve an anticipated effect on this whole surface, a single drop will not suffice, rather we will have to project several bombs toward the same target; these will distribute themselves over the surrounding surface according to the laws of chance. The distribution of hits inside the area will not be uniform; the bombs will strike more frequently in the neighborhood of the target than far away; there will also be unavoidable bomb-hits far outside the area being attacked. However, on the basis of laws of probability, the bomb distribution can be predicted well enough so that the goal of the attack can be achieved with the same or even greater accuracy than for point attack." Sanger clearly had great plans for his fleet of rocket bombers. He suggested that the largest single bomb would be approximately thirty-three feet long, the smallest about ten.

In any event, wherever the massive bomb fell, the destruction would have been impressive—8,000 pounds of high explosive is a major destructive force. Dropping from a high altitude would increase the energy imparted to the target, he reasoned. "Entirely new conditions occur for the area bomb, which has a velocity of impact 10 times as great. The energy of impact is much greater than the energy content of the explosives in the bomb. The strength of the material of the bomb itself will permit it to penetrate a structure, or even to go through a city with numerous buildings, because of the small angle of impact; it will not permit penetration into the earth." In other words, a glancing explosion at the surface of a city would cause a huge, destructive shock wave.

The psychological impact of either kind of bombing would have been immense. By the time serious planning was underway, the Doolittle raid on Tokyo was a couple of years old, but still remembered by the Empire of Japan as a serious blow to their sense of security.[9] That nation had thought itself safe from Allied attack on home soil, and the relatively small number of Doolittle's bombs that succeeded in hitting their targets had left a huge mental scar. An 8,000-pound weapon falling on Manhattan would yield an explosive equivalent in excess of sixteen of Doolittle's 500-pound bombs, and would have had a huge impact on the American psyche, long secure in the knowledge that the war was being fought far from home.[10]

A sign of the optimism that seemed to infect Sanger through 1944 can be seen in his targeting suggestions, for which he wrote, "From the characteristics given for the rocket bomber it follows that this is not the development of an improved military craft, which will gradually replace present types, but rather that a problem has been solved for which no solution existed up to now, namely, bombardment and bombing over distances of 1,000 to 20,000 km. With a single rocket bomber point attacks can be made, from Central Europe, on distant point targets like a warship on the high seas, a canal lock; even a single man in the other hemisphere can be fired upon. . . ." With very limited abilities to target at these speeds and altitudes, however, smiting President Roosevelt might have been a challenge. Sanger continues, "With a group of 100 rocket bombers, surfaces of the size of a large city at arbitrary places on the earth's surface can be completely destroyed in a few days." The Fuhrer must have wept tears of joy when he read this, though he was evidently not sufficiently moved to fund the construction of even one *Silverbird*.

This was just one scenario—there were also more technologically advanced designs for bringing destruction to America. From 1939 onward, the Nazis had been working on nuclear fission with the ultimate goal of building atomic weapons. It was a vastly smaller undertaking than the Manhattan Project in the United States, but by the end of the war it had made significant progress toward creating enough fissionable material to create an atom bomb. If this device had been made available to a flight-ready *Silverbird*, the results would have been quite more dramatic than even an 8,000-pound mega bomb. Had Manhattan been successfully targeted, even a Hiroshima-sized bomb would have devastated much of the city. It was an unfulfilled dream, but a terrifying one.

But with the German war machine being ground down and Hitler's iron control of the campaign becoming increasingly irrational, the military's resources were being rapidly consumed. Their atomic bomb effort was comparatively small, and the *Silverbird* never made it beyond the mock-up stage. There would be no surprise attack on New York City.

And then, there was Wernher von Braun, who was moving forward with his successful V-2 ballistic missile project. An intercontinental version of the V-2 (its technical designation was A-4) called the A-9/A-10 was in the planning stages by 1940, and would have been piloted (the pilot would set an upper stage into a glide toward its target and eject to safety or imprisonment). Had atomic weapons been available, and small enough to be carried by von Braun's rockets, the results could have been sufficient to turn the war to Germany's advantage, at least temporarily. With von Braun's rockets already well in development, the *Silverbird* would have been a huge undertaking to simply add a few more American targets to the list.

The overriding question is, would the *Silverbird* have worked? Most modern engineers who have looked at Sanger's data doubt that the project would have been successful.[11] Such technologically advanced high-altitude flight projects usually involve much more complexity and effort than initially envisioned, and the rocket plane was at the bleeding edge of the engineering know-how of the period. For one thing, the temperatures encountered during the repeated entries into the atmosphere would have been extremely high for the metals available to the German engineers. The alloy used to construct the X-15 more than a decade later was just being developed in England in the 1940s, and turned out to be challenging to fabricate into complex shapes. Welding was particularly difficult, as the alloy, called Inconel, tended to crack during the process. All these challenges were overcome in time, but time was something that Germany had little of by 1944.

Alternatively, they might have developed ceramic heat-resistant tiles such as the ones used to insulate the hull of the space shuttle, but even in the 1970s perfecting those tiles was a long and difficult process that continued to be an issue throughout the program—they were delicate, easily damaged, and each one was custom made to fit. It took a long time to get it right. Again, in a peacetime economy with time to spare, perhaps it could have been done in the 1940s. But in wartime Germany, in the dark corner into which the country was sliding, there would have been little chance of success.

As for rocket power, the ambitious numbers Sanger claimed in his paper look impressive even by today's standards. It would have been nearly a miracle to reach these goals in 1944, especially for something intended to be reusable. And the power output of the rocket engines was connected to the question of effective range. To reach America, and have sufficient range to fly cross-country and past its western border into friendly territory, a launch site in the Azores, far off the coast of Portugal, was suggested. This would not only dramatically decrease the range the *Silverbird* would have to fly—the archipelago was 850 miles west of Continental Europe—but was also closer to the equator, a preferable launch site to the other northern European latitudes that Germany had access to. However, this launching facility would have been an irresistible target to Allied bombers, and difficult to defend from air attack. Furthermore, in anticipation of German activity there, the United States had contingency plans for an assault on the islands by the US Marine Corps. Given the ferocity with which that force later took control of well-defended islands within the Japanese Empire, it is unlikely that a German rocket base would have survived for long.

Finally, it is unlikely that the flight rate of the *Silverbird* would have been anything like Sanger envisioned. Witness again the space shuttle, which at one point was thought to be capable of up to fifty-four flights per year in peacetime and with 1970s technology—it ultimately struggled to average more than nine flights per year.[12] While the shuttle was technologically more complex, it is unlikely that the *Silverbird*'s turnaround requirements, being supported by a struggling German military far away on the mainland, could have been met. Compellingly, in his 1944 study, Sanger mentions reusability and the economies resulting from it more than once, but in practice, the outcome would likely have been far less appealing.

Still, had the *Raketenbomber* worked, the outcome of WWII could have been significantly altered. While it is thought to be unlikely that the US would have ever surrendered—the raw productivity of wartime America would eventually have stopped the attacks—the war could have lasted much longer or, less likely, resulted in a negotiated peace. The authors of endless alternative history fiction have much to say on the subject, but real answers are impossible—it is simply far too speculative. But a space-faring Nazi regime would have been something to inspire fear in anyone exposed to open skies.

CHAPTER 2

RED MOON: COUNTERING THE COMMUNIST THREAT ON EARTH AND IN SPACE

CLASSIFIED: *DECLASSIFIED IN 2014*

> *"The employment of moon-based weapons systems against earth or space targets may prove to be feasible and desirable. . . ."*
> **—Project Horizon Report: Volume I,**
> United States Army, June 9, 1959

In the opening days of the space age, paranoia had a large role in planning for the "conquest of space," as the media often referred to the struggle for dominance in the heavens. While the closing salvos of World War II were more than a decade past, the effects of the massive conflict continued to reverberate throughout the world. The defeat of Germany and Japan by the Allied powers was conclusive, leaving both those countries in smoking ruin. The end of the war for Japan had particular significance, as the deafening twin percussive notes of two atom bombs concluded that country's wartime ambitions. But when that horror was unleashed, the nuclear genie was out of the bottle.

So while the former Axis powers rebuilt their nations from smashed and charred rubble, on the far side of ruined Germany, the Soviet Union also restored order to its society. While an uncomfortable alliance between Russia and the US had held steady through the course of the war, relations took a bitter turn after 1945. The United States and the USSR emerged as the two most powerful nations from wartime, and within short order became enemies, each distrustful of the other's intentions.

The worldviews and ambitions of the two nations were simply incompatible. The tensions created by this new world order would result in some bizarre and ultimately impractical schemes. But impracticality didn't stop the militaries of each nation from trying.

AIR LOCK & LIVING QUARTERS

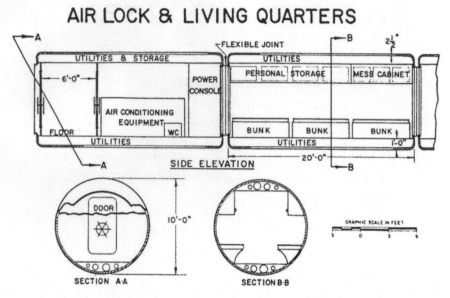

Fig. 2.1. The Horizon base would be constructed of cylindrical modules, launchable on large boosters. Some would be for cargo, and some would be habitats, as seen here. Image from DOD/US Army.

For some time, the US held a military advantage, having developed and used the terrible destructive power of the atom bomb, which, if delivered to any populated area, assured massive destruction and loss of life. As the world slowly recovered from the rigors of war, an uneasy peace reigned, with one nation holding the nuclear mallet quietly aloft, and all the power this implied.

Then, in 1949, the balance tipped. Through intensive espionage efforts dating back to the war, the Soviet Union managed to obtain American experimental data on atomic fission and atom bomb designs. In August of that year, the first Soviet A-bomb test shocked Western governments and resulted in an immense boost of the USSR's military prestige. Besides a massive postwar conventional military machine, the Soviets now had atomic weapons as well. The Cold War had begun—the longest peaceful standoff of antagonistic world powers in history, each of which would soon have the technology to virtually annihilate the other within hours.

At the time, the only reliable method of delivering a bomb to the enemy was the airborne bomber. The US continued to develop aircraft along the lines of their Japan-smashing B-29 Superfortress, a massive four-engined airplane that could carry a nuclear weapon well over 3,000 miles. The Soviet Union had made great strides in its postwar air forces, and soon had planes of similar capacity and ability. But formations of large, lumbering bomber aircraft were unwieldy and vulnerable—they could be intercepted and shot down—so a better method of delivering city-incinerating payloads was deemed crucial.

As the two countries continued to build massive bomber fleets, defensive fighter aircraft, radar networks, and surface-to-air missiles, something more exotic crept into the mix. And, not without irony, the technological roots came from a mutual enemy of the former allies—Germany.

Prominent in the group of impressive minds was Wernher von Braun. Born into the German aristocracy in 1912, he would turn out to be a pivotal persona in rocket development in both Germany and, later, in the US. He was an unremarkable student in his youth, but his passion for spaceflight was ignited by reading the same book that had inspired Eugen Sanger, Oberth's *By Rocket into Planetary Space*. In this work, von Braun found his purpose, and by 1930 he had earned a degree in mechanical engineering and in 1934 a PhD in physics. While in school he joined the German Spaceflight Society and was soon joining his fellow engineers and spaceflight devotees in building small liquid rockets.[1]

As Germany drifted toward WWII, von Braun joined the Nazi party in 1937 and the SS, the much-feared German military elite, in 1940. Von Braun later insisted that both these acts were the results of pressure brought to bear by the German government,[2] but this is widely disputed. Most historians feel that his membership in both organizations was a pragmatic decision designed to allow him to further his work with rocketry under Nazi auspices, since once the Nazis were in power, work in amateur rocketry was forbidden. In any event, his knowledge of rocketry was noted by the military as early as 1932 when he began working in coordination with the German army.

By the time America entered the war, von Braun was in charge of a large team of engineers, scientists, and other experts designing and building ballistic missiles for the German war effort. His V-2 was the culmination of these efforts. While primitive by today's standards, it was an exotic piece of machinery in an age that knew little of liquid-fueled rocket engines, much less ballistic missiles. By the end of the war, V-2s had pummeled numerous targets in Europe. With its primitive guidance system, the V-2s were undiscriminating beyond their ability to target a specific city up to 500 miles away with some success. In the end, they made little difference in the German war effort, but the scene was set for the further development of warhead-bearing missiles and the space age.

As WWII was nearing its conclusion, von Braun knew he would have to surrender to the Americans or the Soviets, but whichever side he chose, he would have to proceed with care. The SS was guarding him and his colleagues, and would likely exterminate the lot of them to prevent rocket secrets from falling into Allied hands. Von Braun chose the United States for a number of reasons, not the least of which was the Red Army's reputation for seeking revenge, and he soon found himself and a select group of his associates at Fort Bliss in Texas, and later in White Sands, New Mexico. From this second, rather rural posting, he continued to develop rocket technology with captured V-2s. He fired them off with some regularity, further refining his designs.[3]

After a few years in the windswept confines of White Sands, he was finally transferred to the Army Ballistic Missile Agency in Huntsville, Alabama, with the Cold War dictating the course of rocket development. By 1952, von Braun was leading the effort to develop a short range ballistic missile—an SRBM, the forerunner of the modern intercontinental ballistic missile (ICBM).[4] The Soviet Union had been developing missiles intended to complement, and eventually surpass, their own nuclear bomber fleet, and the US needed to keep pace. The Russians were good at rocketry (having captured their own Germans, V-2 rockets, and plans at the end of the war), and had rapidly pulled ahead of the US in raw rocket power and reliability. The vast inventory of American bombers and escort aircraft would be of little deterrence if Russian nuclear missiles (which could not be intercepted) could arc overhead and wipe out the United States while the US aircraft were still lumbering to their targets in Russia.

Von Braun's new rocket was called the Redstone, and it was at its core a redesigned and improved V-2. This was America's first missile of any size, and was used for early tests of rocket-propelled nuclear weapons. Von Braun further refined his rocket designs, using more advanced fuels and technologies for ever-increasing power. The family of rockets that descended from the Redstone included the Jupiter, the Juno, and ultimately the Saturn series, the final version of which launched the Apollo program to the moon.[5]

Then, on October 4, 1957, came an event that electrified the world. The Soviet Union used one of its powerful R-7 missiles to launch the world's first manmade object into orbit. Called *Sputnik*, it was little more than a hollow metal ball with batteries, a radio transmitter, and some rudimentary sensing equipment aboard. But the little spacecraft's regular beeps could be received by listening posts around the world, and that tiny radio was, in propaganda terms, more powerful than a thermonuclear warhead. The Western press trumpeted the story across the headlines—"SOVIET FIRES EARTH SATELLITE INTO SPACE; IT IS CIRCLING THE GLOBE AT 18,000 MPH . . ." and in smaller print, the stinger: "Device is 8 Times Heavier Than One Planned by the US."[6] That last bit stung. Both the US and USSR had been engaged in trying to get a satellite aloft, but the Americans had entrusted the project to the US Navy and its tiny Vanguard rocket was plagued with troubles and, in any case, would be unable to orbit anything much larger than a grapefruit. Sputnik was twenty-two inches in diameter and weighed 184 pounds. A month later the Soviets launched Sputnik 2. This second satellite weighed almost *1,200 pounds* and carried a live dog into orbit (though she did not live for long). The implications of the rapid increase in carrying capacity was not lost to the US military and many civilians—the Russian rockets would be able to carry nukes through space to fall at transonic speeds into US territory, and, unlike bombers, would be virtually unstoppable.

On December 6, the US Navy hoisted the Vanguard TV-3 rocket up to its launch gantry at Cape Canaveral. The press had been invited, and the cameras were rolling—this would be America's answer to the Red Threat. At T-zero the rocket's engines ignited and it lifted off the launchpad—to a height of about four feet. Then, while still a few hundred miles short of its goal, the rocket lost thrust and came right back down, crumpling and exploding in a huge orange fireball. If this failure was not sufficiently embarrassing, the Vanguard satellite, that miniscule, melon-sized metal sphere, toppled off the top of the rocket and rolled to battered safety. It was later allegedly found near a Dumpster not far from the now-scorched launchpad, its radio transmitter still faithfully beeping away. It was hustled off for examination and ended up in a museum.[7]

The Soviets had scored two public relations coups, and the United States was mired in failure. The newspapers called it "Flopnik," "Kaputnik," and other unflattering names.[8] The navy went back to its workshops and did eventually succeed in getting Vanguard into orbit, but President Eisenhower had lost patience with the over-schedule and underperforming program, and wisely turned to von Braun.

There had been reluctance to work with the brilliant German engineer—the key word was "German." While the American government had done some work to cleanse the public image of him and his fellow rocketeers, the memories of London, Antwerp, and other European capitals in flaming ruin at his hands were recent enough that many had preferred a success engineered by "true" Americans. But with the failure of Vanguard, there was no time for such niceties. The administration asked von Braun if he could get a satellite up in short order.

Von Braun had been waiting for such a moment. He had seethed under the yoke of the army at Huntsville, relegated to developing nuclear missile nosecones and boosters, but absolutely not allowed to pursue anything that was specifically intended to put an object into space, which had been his passionately held goal since at least 1930. Now was his chance to shine. "We can fire a satellite into orbit in 60 days from the moment you give us the green light," he said with confidence.[9] The startled secretary of defense, Neil McElroy, suggested that ninety days might be more realistic. In the end it took eighty-four.

On February 1, 1958, the thirty-pound satellite, Explorer 1, flew into orbit atop one of von Braun's Juno rockets. It was small, but it was a satellite, and carried some science instrumentation that resulted in the discovery of the Van Allen radiation belts surrounding the Earth. However, the Soviets still had a huge lead, with big rockets carrying heavy payloads into space that embarrassed US efforts. And those big Russian rockets could clearly carry nuclear weapons. Something had to be done to counter this perceived threat, and quickly.

It was in this atmosphere of technological and military paranoia that the US Army initiated a study called Project Horizon to determine the viability and utility of a military moon base. Fresh off his rescue of the US satellite efforts a year before, von Braun was brought in to consult. Initiated in March 1959, the Horizon study resulted in a detailed report just three months later. In brief: the US Army could have a lunar base, housing twelve battle-ready US Army soldiers, able to spy on any part of Earth and rain nuclear death on our enemies, by 1966.

The meaning of the proposal was made clear at the outset: "Moon-based military power will be a strong deterrent to war because of the extreme difficulty, from the enemy point of view, of eliminating our ability to retaliate."[10] By placing nuclear weapons on the moon with the ability to reach anywhere on Earth, it was thought that the US would have a staggering strategic advantage.

Never mind that it would take those weapons two days to cross the distance between the two worlds, or that ICBMs would shortly do the job just as well within minutes. Viewed from the perspective of 1959, the general concept was not as zany as it first sounds. At the time, missile guidance systems and flight robotics were still in their infancy. Earth-observation satellites such as those fielded in the 1960s, which could deliver images of Soviet territory within a few days of launch, were non-existent. There were no submarine-launched ICBMs, no instant high-altitude overview of enemy activity. So space was the classic "high ground" armies have sought throughout history, and human operators would be needed to carry out the tasks we now regard as being routinely robotic and, in many cases, autonomous. So from a human-operated surveillance and deterrence perspective, Horizon made some sense.

On the other hand, the Horizon study was first envisioned just a year after the launch of Explorer 1—the US had only a tiny bit of spaceflight experience, and that with a tiny, crude satellite. Most Americans still did not have color televisions, and home electronics, such as they were, used vacuum tubes in lieu of the transistor. Computers were a mere shadow of what we had just a few years later, and most places using them leased mainframes from IBM or other companies for vast sums. In early 1959, no human had yet flown higher than about 100,000 feet, or eighteen miles, and that was done in balloons (the X-15 would not fly for a few months), and the boundary to space was agreed to be about sixty-two miles in altitude. So you can see that the scope of Project Horizon involved a bit of hubris on the part of its designers. But this is how dreams are made into realities—through aggressive and optimistic goals that become whittled down and accomplished over time, usually at budgets well beyond those initially calculated. Horizon would certainly have fallen into this budgetary category had it been attempted.

Everything about Project Horizon was big. The goals, as laid out in an opening statement in volume 1 of the study were ambitious: "There is a requirement for a manned military outpost on the moon. The lunar outpost is required to develop and protect potential United States interests on the moon; to develop techniques in moon-based surveillance of the earth and space, in communications relay, and in operations on the surface of the moon; to serve as a base for exploration of the moon, for further exploration into space and for military operations on the moon if required; and to support scientific investigations on the moon."[11]

Of particular interest to the military was, of course, the idea of continuous observation of the USSR. The notion of claiming the moon as de facto territory was a strong motivator as well; no treaties regarding the territorial claims in space would be signed until the 1960s. So, while science and exploration were mentioned, these were not primary goals—militarization was.

"The employment of moon-based weapons systems against earth or space targets may prove to be feasible and desirable," the study continues. "Any military operations on the moon will be difficult to counter by the enemy because of the difficulty of his reaching the moon, if our forces are already present and have means of countering a landing or of neutralizing any hostile force that has landed. The situation is reversed if hostile forces are permitted to arrive first. They can militarily counter our landings and attempt to deny us politically the use of their property."

Science also got its due, and with some prescience: "The scientific advantages are equally difficult to predict but are highly promising. Study of the universe, of the moon, and of the space environment will all be aided by scientific effort on the moon. Perhaps the most promising scientific advantage is the usefulness of a moon base for further explorations into space. *Materials on the moon itself may prove to be valuable and commercially exploitable.*"[12]

The whole enterprise depended on von Braun's designs for the Saturn family of rockets. Most of us know the Saturn V as the mammoth rocket that flew NASA's astronauts to the moon, but that is an entirely different machine from the early Saturn boosters. These were based on the Redstone, clustering components of that rocket to combine their thrust into one larger machine. The total output of the Saturn 1 was 1,500,000 pounds, vastly larger than any American rocket of the time, but only equivalent to one of the five Saturn V's first-stage rocket engines that would power the Apollo flights just a decade later. But it was what was available.[13]

Up to 229 Saturn 1 rockets would be required to assemble the components needed for the Horizon moon base and the associated orbital machinery; in reality only nineteen of the early designs of the Saturn were ever flown,[14] being superseded by the mammoth Saturn V (which itself flew a total of thirteen missions, eleven with a crew). So the rocket hardware requirements were, in a word, *massive*.

The timeline went like this:

1963: The Saturn 1 rocket would be fully operational and used to build a small space station, composed primarily of spent upper-stage fuel tanks. This station would be used for refueling operations and other logistics in Earth orbit.

1964: A second orbital station would be constructed, and one of the two would be outfitted with life-support capability to house a crew. Lunar landing and ascent vehicles would be tested and readied to begin the task of hauling men and materiel to the lunar surface.

1965: Cargo flights to the moon would begin. Two astronauts would arrive at the moon to explore the environment and prepare for further crews to arrive. Later in the same year, a crew of nine more astronauts would arrive to take over the construction of the base, estimated to take eighteen months. Tractors, cranes, and other needed construction hardware would be assembled. The first habitation module would be outfitted, and two previously delivered nuclear reactors (later expanded to four) would be buried and activated to provide power. More habitat modules would be outfitted, and spent rocket stages would be used to house machines and supplies.

1966: The habitat modules would be covered with lunar soil to protect against the radiation expected in the lunar environment (though little was known at the time about how intense this might be). The complement of twelve crewmen would arrive and take up nine-month rotations at the outpost. A total of 252 men would have departed Earth, forty-two of whom would have continued on to the moon. Moon base crews could eventually reach twenty or more men.

As the study put it: "Initially the outpost will be of sufficient size and contain sufficient equipment to permit the survival and moderate constructive activity of a minimum number of personnel (about 10–20) on a sustained basis. It must be designed for expansion of facilities, resupply, and rotation of personnel to insure maximum extension of sustained occupancy."

The Saturn rockets would have been flying at a rate of 5.3 per month during the buildup phase, for a total of 149 rocket launches, moving 245 tons of cargo and machinery into space. Up to sixty-four more launches were estimated to be necessary for the first year of operations, moving another 133 tons of cargo. Finally, sixteen Saturns would be kept on standby for "emergency" purposes (a number nearly as large as the total actually launched by 1975).

The budget for this undertaking was estimated at about $6 billion. Since the Apollo program, which was a vastly less complex and smaller undertaking, cost over $20 billion by the time it was complete,[15] it's clear that the scope of development costs for Horizon were vastly underestimated. But that's how space works—everything is harder than it looks, a reality that continues to rear its ugly head even today—witness the vast cost overruns of projects such as the space shuttle, the International Space Station, the Mars Science Laboratory, and the James Webb Space Telescope—all of which cost much more than originally planned.

But all this overlooks one point—the almost primal drive to militarize the moon. While I've mentioned the "high ground" argument, let's look a bit deeper at its implementation. A quote from one of Horizon's leading advocates, Army General John Bruce Mendaris, is instructive:

"The moon has no water, it has no air, but it has land. That's the army's chance."[16]

Clearly this was a tactical endeavor as much, if not more so, than a scientific one. The base was designed to be defensible from overland attack—the new US "space army" was thinking an awful lot like the traditional ground-pounding army. Two principal weapons were to be fielded to the crew members—lunar soldiers, in effect—while on assignment to Horizon.

The first weapon system was a revised version of the army's Claymore mine. The existing M18 Claymore was (and is still) a mine designed to be used aboveground, on a stand, or attached to a tree—its case is curved to the shape of a medium-sized tree trunk. It is classified as an "anti-personnel" mine, meaning that it is specifically designed to kill people, not to disable machinery. It fires a broad swath of metal pellets, like a huge shotgun blast, in the direction it is facing (the front of the case says, "Front Toward Enemy"[17] lest there be any confusion, which could have unfortunate results). A new version of the M18 was suggested for development specific to Horizon's defensive needs—puncturing space suits. The pellets didn't have to kill the enemy moon soldier; the vacuum of space would do that.

A more bizarre weapon would also be deployed on the moon, one being developed by the army for use in the European theater of the 1950s and 1960s. The M-29 Davy Crockett Tactical Nuclear Recoilless Gun was suggested as additional armament for the Horizon lunar base. This was a small rocket launcher that could be set up by a few soldiers on a small tripod, and which fired an M388 nuclear warhead—a fifty-one-pound projectile that had the equivalent punch of ten to twenty tons of TNT. The idea was that this weapon system could be transported in the lunar field and set up quickly for local defense against overland troops assailing Horizon from the surrounding territory—presumably Soviet soldiers, as there was nobody else on Earth who had the technology at hand that could even begin to approach what would be needed to engage in this kind of off-Earth combat.

Additional smaller weapons for use in a lunar environment were later considered, and are discussed in chapter 20.

Today, it's hard to take seriously the idea of Red Army lunar marines assaulting Horizon Base One overland, sidestepping crater ridges and bounding along in one-sixth of Earth's gravity while brandishing their vacuum-capable rifles and (likely) nuclear bazookas of their own. But at the time, the threat must have seemed deadly real to members of the American military establishment who saw an ever-ascendant USSR that was leading the way in large rockets and heavy orbital satellites. If they could put a dog in space, why not men on the moon?

In conclusion, the report said, "There are no known technical barriers to the establishment of a manned installation on the moon." That was a bit of an understatement. While there may not have been any *conceptual* impossibilities in the report, the sheer scope and the many intense technical challenges were obvious. Project Horizon was shelved in short order.

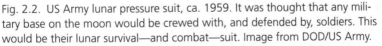

Fig. 2.2. US Army lunar pressure suit, ca. 1959. It was thought that any military base on the moon would be crewed with, and defended by, soldiers. This would be their lunar survival—and combat—suit. Image from DOD/US Army.

As we came to understand over the next ten years, the decade between the Horizon proposal and the first footsteps on the moon in 1969, spaceflight is a lot harder than it looks. If today you make the trek to the Space Center Houston museum in Texas, or the Kennedy Space Center museum in Florida, you can look up at one of the remaining Saturn V rockets left over from the Apollo program (Apollos 18, 19, and 20 were planned and built but ultimately cut from the program, leaving behind some of the most expensive museum exhibits in the world). As you gape in awe at the 363-foot-long, 6.5-million-pound behemoth, stroll from the five massive fuel-guzzling F-1 engines on the business end of the rocket toward the pointy end. That little tiny capsule is all that returned from the lunar voyage, and it took that enormous machine just to get three men, sandwiched inside the capsule, to the moon and back (the lunar modules were left there, with the exception of Apollo 13's). That tiny spacecraft, loaded with fuel and equipment and astronauts and its lunar lander, weighed about fifty tons. You can imagine what it would have required to get a full lunar base, moon tractors, rotating crews of ten to twenty men, food, weapons, oxygen, water, and of course nuclear bazookas to the moon.

And what of the Soviet threat to dominate space and create a Red moon? When President Kennedy kicked off the American lunar landing effort in 1961, with only fifteen minutes of manned[18] spaceflight time under the US's belt (via Alan Shepard's suborbital Mercury flight), the Soviet Union, vastly more experienced in space, began its own lunar landing program. But despite Herculean efforts, and a family of Soviet lunar-capable hardware developed and partially tested, the USSR never got close to making a successful landing on the moon with its single-cosmonaut lander. In fact, the Soviets never sent a cosmonaut beyond Earth orbit. Each of their huge N-1 moon rockets failed during flight testing, and the program was abandoned in 1976.[19] In the end, their ability to build a lunar base to threaten US interests on the moon was nil, less likely even than a successful execution of Horizon.

President Eisenhower, Kennedy's predecessor, was never keen on the military dominating American space efforts, much to the surprise of some. He insisted on the formation of a civilian space agency, and NASA was founded in July 1958. Von Braun and his team were absorbed into NASA. The US Army would continue to fight ground wars for its country, and NASA would ultimately win the Cold War in space, with those first tentative steps on the moon on July 20, 1969.

CHAPTER 3

DAS MARSPROJEKT: RED PLANET ARMADA

UNCLASSIFIED

L ots of people have dreamt of going to Mars over the last century. Elon Musk wants to go (and is willing to pay hundreds of millions to get there). Buzz Aldrin would like to fly there in a spacecraft of his own design. Tom Hanks has raised his hand, and Carl Sagan famously expressed interest. And of course, as you might expect, Wernher von Braun would have been thrilled to lead an expedition.

Von Braun's interest was more formal than most, and prompted him to write two books about how a voyage to Mars might be accomplished. The first was a novel that went unpublished for many years, but the second was a short treatise that included a mission plan, spacecraft designs, trajectories, fuel loads, calculations, and more. It was called *Das Marsprojekt* when published in German in 1952, and *The Mars Project* when published in English in 1953.[1]

This was no mere flight of fancy. Von Braun broke out the slide rule and pencil and did the nuts-and-bolts calculations to craft a workable mission plan. It was as detailed and comprehensive as Project Horizon was vague and uncertain. Given the knowledge of Mars and its environment in 1948, in was an ambitious undertaking, especially considering that von Braun was working from data on Mars that had advanced only incrementally since the early twentieth century. Mariner 4, the first spacecraft to survey the red planet, would not reach Mars until 1965—sixteen years after the book's first publication. The data received from the spacecraft indicated a planet far more inhospitable than previously thought. So, in hindsight, the mission plan outlined in *Das Marsprojekt* would not have worked in a few critical ways—including von Braun's ambitious plans to land via winged gliders—but was not as far off as one might think.

A word here about the perceptions of Mars in the 1940s and 1950s. From the time of early telescopes, astronomers had been studying the planets, and Mars, being our nearest neighbor after the moon and Venus, was of great interest. The moon was a fine telescopic target, but was quickly understood to be a lifeless world—it has no atmosphere, and the surface temperatures range from a scorching 253°F to a chilly –243°F. There is little there but rock and dust, mountains and craters. The moon was fascinating, but not a place that might accommodate human visitors easily.

Then there was Venus. The planet is perpetually covered by dense clouds, so optical telescopes could accomplish little beyond studying the cloud tops and doing spectroscopic studies of the atmosphere. Venus had long been thought to possibly be a wet, tropical world—that its thick, carbon dioxide–dominated atmosphere would promote riotous growths of swamps and tropical forests, possibly with living creatures. But robotic probes began studying Venus about the same time as Mars, and it was far beyond tropical. The clouds that shroud the planet are sopping with sulfuric acid, and the temperature at its surface is over 800°F. The terrain looks like a shale-covered desert lit by lightning storms and drenched with acid rain.

But Mars intrigued. The planet was close enough to be observed through even small telescopes, and larger instruments—say, twenty-four inches in diameter and above—could resolve fuzzy details on the surface. Such a telescope was central to the efforts of amateur astronomer Percival Lowell, an avid observer of the Martian surface, who commissioned a purpose-built observatory for investigating Mars. The observatory, located atop a mountain in the Arizona Territory, was completed in 1894.[2]

Lowell's fascination with Mars had been ignited by the work of a previous generation of astronomers such as Giovanni Schiaparelli. Working with a small refracting telescope in an observatory at the Brera Observatory in Milan, Italy, Schiaparelli drew remarkably detailed maps of Mars by observing the planet night after night, for entire seasons, when it was at its closest approach to Earth (this occurs every two years). He would combine these sketches into more detailed drawings, ultimately creating his exquisite maps. Unfortunately, the telescope he used was a mere eight-inch-diameter refracting instrument, which was not very big for observing far-off planets.[3] Its limited resolution, combined with the ever-present interference of Earth's atmosphere, produced little more than a large and wavering red blob with some lighter and darker shading scattered across its surface. So when Schiaparelli combined his sketches of the fuzzy images, a bit of imagination was bound to "enhance" the process. On his maps, the planet went from a hazy disk to a detailed globe, covered with surface features, which were connected by lines that he called "*canali*," the Italian word for "channels." It has also been suggested that, due to the lack of modern non-reflective coatings on the optics of the time, he may have even been charting the blood vessels of his own retina, reflected back by a lens in the eyepieces he used. Whatever the cause, his maps were incredible works of draftsmanship, art, and imagination—a combination that was to cause a furor in short order.

When he published some of his best work, a compilation of observations between 1877 and 1886, Schiaparelli's *canali* caused the most interest, as the English-speaking world quickly interpreted this to mean "canals." "Channels" can occur in nature—the result of water runoff or faulting—but canals are *built*, presumably by intelligent beings. The media picked up on this, and thus Mars was reborn as a world with inhabitants intelligent enough to build a system of irrigation canals.

By the time Lowell began researching Mars in his custom-built observatory, the idea of an advanced civilization on Mars had become increasingly hard to ignore, and he bought into it completely. Mind you, he was no scientific slouch—Lowell held a degree in mathematics from Harvard. But his passions for Mars, and the possibility of intelligent beings inhabiting the planet, drove his thinking, and it was in this frame of mind that he inaugurated his new observatory. Optical observations were the best tool available for researching the Martian environment, and his new 24-inch telescope was one of the best available (the only alternatives to examination by the eye were limited telescopic photography and the spectrograph, an instrument that breaks up light into individual chemical "lines," allowing the examination of specific chemical elements in the atmosphere of the object being observed). Lowell took to his task with a passion, observing Mars for months at a time and creating his own improved maps. And what maps they were—even more intricate than Schiaparelli's, with each line, dot, and smudge numbered, labeled, and described. As his maps took form, his fertile imagination also took flight, and he was soon writing what would become a series of best-selling popular books about Mars. And these were not dull, scientific treatises of surface markings and chemical tables; the books were filled with descriptions of both the physical Martian environment and a world inhabited with intelligent beings and advanced engineering.

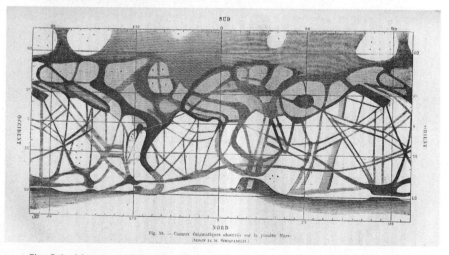

Fig. 3.1. Mars as envisioned at the time *Das Marsprojekt* was written. This map, created by astronomer Giovanni Schiaparelli around 1886, was in wide use until the first images came back from Mariner 4 in 1965. When von Braun was planning his Mars expedition, little was known about conditions on the planet, in particular the density of the atmosphere—it was thought to be much higher than it actually was. His winged gliders would have crashed had they attempted a landing. Image from Wikipedia.

Lowell's Martian empire came to rival that created soon after by the novelist Edgar Rice Burroughs.[4] And while Lowell did not have Burroughs's sword-wielding, six-limbed warriors riding multi-legged *thoats* (a kind of mutated Martian horse) ranging across the planet's surface, he did invent an advanced civilization, complete with a suggested form of government, impressive technology, and, most importantly, a reason for constructing the vast, planet-girdling canals he thought he saw.

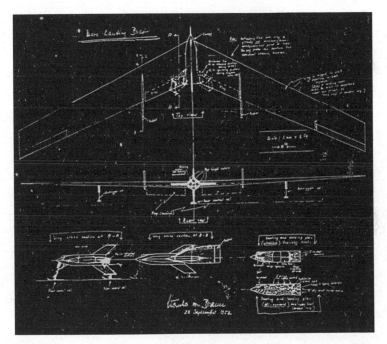

Fig. 3.2. A sketch by Wernher von Braun depicting his design for large gliders to land humans on Mars. The first would land at the Martian pole carrying a tractor and supplies, and the crew would then drive 4,000 miles toward the equator to prepare a runway for additional landers. It was an audacious and, as it turned out, impossible plan. Image from NASA.

It should be noted that, within academia and "professional" circles, Lowell (despite his Ivy League degree) was considered by many to be an *amateur* (often said with quiet distaste), despite rigorous self-training and his fine observatory. He was in an unusual position within the astronomical profession—too imaginative to be fully accepted, but too rich and well equipped to be ignored. Regardless of the perceptions of the scientific elite, for nearly two decades he made careful and detailed studies of Mars through his large telescope, resulting in reams of sketches, many splendid maps, detailed globes, and, most influentially, multiple books and papers, including *Mars* (1895), *Mars and Its Canals* (1906), and *Mars as the Abode of Life* (1908). The articles were widely published, the books sold well, and both left new and fantastic impressions in their broad readership—while many scientists sniffed, the *hoi-polloi* ate it up.

Lowell reasoned that the planet had followed a different evolutionary path from Earth (and he was right on that score), and had begun to wither away much sooner than its sister planet. The planet's water was drying up (as indeed it did, but billions of years before he thought), and the ever-practical Martians used their unified global government to initiate huge engineering projects to construct canals, some thousands of miles long, to channel water from the icy polar caps to the thirsty equatorial regions, in order to survive.

He went so far as to predict what kinds of technology the Martians must have invented over the millennia, being a race far older than our own. Television, advanced transportation machinery, and the massive hardware required for planetary excavation projects all sprang forth from his fevered imagination. He reasoned that to accomplish these vast feats of global engineering, the Martians must be a peace-loving civilization and were far advanced beyond his perpetually squabbling brotherhood of mankind. The citizens of the red planet were perpetually bonded by the ever-present struggle to survive in a harsh, dying ecosystem.

It should be noted that Lowell was one of the few "intelligent Martians" promoters who utilized reasonably solid scientific reasoning to come to his conclusions. If you were to read his books today, bearing in mind the sketchy evidence available to him and others of the time, and can survive the Victorian pacing of the flowery descriptions, you can enjoy the roadmap of his logic and feel the genuine passion of his ideas. He was no tabloid journalist, despite the fantastic thoughts he put forth. In his 1895 best-seller *Mars*, he wrote the following of the infamous canals:

[T]he aspect of the lines is enough to put to rest all the theories of purely natural causation that have so far been advanced to account for them. This negation is to be found in the supernaturally regular appearance of the system, upon three distinct counts: first, the straightness of the lines; second, their individually uniform width; and, third, their systematic radiation from special points. ... Physical processes never, so far as we know, end in producing perfectly regular results, that is, results in which irregularity is not also discernible. Disagreement amid conformity is the inevitable outcome of the many factors simultaneously at work ... That the lines form a system; that, instead of running anywhither, they join certain points to certain others, making thus, not a simple network, but one whose meshes connect centres directly with one another,—is striking at first sight, and loses none of its peculiarity on second thought. For the intrinsic improbability of such a state of things arising from purely natural causes becomes evident on a moment's consideration. ... Their very aspect is such as to defy natural explanation, and to hint that in them we are regarding something other than the outcome of purely natural causes.[5]

You simply must admire anyone who can use the word "anywhither" with authority. Clearly, in his mind, the canals girdling Mars had to stem from artificial, i.e. intelligently constructed, origins. Lowell continued,

Martian folk are possessed of inventions of which we have not dreamed, and with them electrophones and kinetoscopes are things of a bygone past, preserved with veneration in museums as relics of the clumsy contrivances of the simple childhood of their kind.[6]

From Lowell's standpoint, the Martians had been at this a long time and were simply too smart to *not* have these things. To close:

The chain of reasoning by which we have been led to regard it probable that upon the surface of Mars we see the effects of local intelligence. We find, in the first place, that the broad physical conditions of the planet are not antagonistic to some form of life; secondly, that there is an apparent dearth of water upon the planet's surface, and therefore, if beings of sufficient intelligence inhabited it, they would have to resort to irrigation to support life; thirdly, that there turns out to be a network of markings covering the disk precisely counterparting what a system of irrigation would look like; and, lastly, that there is a set of spots placed where we should expect to find the lands thus artificially fertilized, and behaving as such constructed oases should. All this, of course, may be a set of coincidences, signifying nothing; but the probability points the other way. As to details of explanation, any we may adopt will undoubtedly be found, on closer acquaintance, to vary from the actual Martian state of things; for any Martian life must differ markedly from our own.[7]

As Western civilization careened into World War I, public attention was drawn away from the Martian miracle workers and refocused on the abuses of our fellow Earthmen. But after the war, the fascination with Mars was renewed, and Lowell's ideas remained in the popular mindset for over forty years, despite the fact that each succeeding decade brought new observations by other astronomers with spectrographs and the associated cold logic that ate away at Lowell's "intelligent Martians" theories. But nobody could say for certain who, or what, might dwell on that parched world.

It was into this arena that von Braun matured, as he wrestled his ever-larger rockets into technological submission. He seems to have been less concerned with what might be *living* on Mars than with the idea of simply *getting there*. Mars, as well as the moon, held a special place in his heart as a planet that could be—*should* be—reached by men in rocket ships.[8] By the time WWII was over and his captured V-2 missiles were launching from New Mexico (and sometimes landing in Old Mexico—at least one careened out of control, digging a large hole in Mexican territory),[9] Wernher had once again focused his goals on reaching space with humans. He proposed first building an orbiting space station (see chapter 6), then equipping an expeditionary fleet to strike out for Mars. In his copious free time at White Sands, New Mexico, between his arrival in the United States in 1945 and his move to Huntsville, Alabama, in 1950, von Braun put pen to paper to expand on his ruminations about sending manned rockets to other planets. The first result was a manuscript for a novel called *Project Mars: A Technical Tale*, which was roundly rejected by the publishers of the day. The characters were paper thin, and the plot contrived. But the fictional setting was merely a backdrop for his beloved Mars mission anyway, and much of the well-studied technical material made its way into *Das Marsprojekt*, or *The Mars Project*. This book, eighty-one pages in German (ninety-one in English), detailed the engineering behind the mission described in his novel. While not a complete engineering study, which would have been ten times longer or more, it was a solid outline of a mission to Mars.

That is, if Mars had been the world we thought it was in the late 1940s. Unfortunately, when Mariner 4 swept past Mars in 1965, it revealed a much bleaker world than previously imagined, with an atmosphere only about one one-hundredth that of Earth's. Prior estimates had averaged about one-tenth of earthly sea level. The idea of sunning oneself by the arid shores of Schiaparelli Crater with a bit of added oxygen via facemask was forever shattered; the environment was vastly more hostile than presumed. This was just one revelation. The ever-changing areas of dark and light on the surface, so long observed via telescope and, as late as 1964, postulated as vegetation blooming and receding with the Martian seasons, turned out to be planet-girdling dust storms and deadly dry collections of craters, extinct riverbeds, and darker shades of rock and soil.

Of course, these realizations would not stop von Braun from his mission to conquer the cosmos with Mars as a first planetary stopover. But the physics of the thin atmosphere doomed his design for the Martian landing, and other factors, such as the radiation that washes the planet's surface, would pile one challenge after another atop his expedition plans.

The harsh realities of the 1960s aside, *The Mars Project* was still a solid piece of work for its day and bears a closer look. From his introduction to the English version:

Soon after the publication of the first serious papers on space travel, a spate of fanciful materials appeared—purporting to show vividly just how an interplanetary voyage would be carried out. The central figure in these stories was usually the heroic inventor. Surrounded by a little band of faithful followers, he secretly built a mysteriously streamlined space vessel in a remote backyard. Then, at the hour of midnight, he and his crew soared into the solar system to brave untold perils—successfully of course.

Since the actual development of the long-range liquid rocket, it has been apparent that true space travel cannot be attained by any backyard inventor, no matter how ingenious he might be. It can only be achieved by the co-ordinated might of scientists, technicians, and organizers belonging to very nearly every branch of modern science and industry. Astronomers, physicians, mathematicians, engineers, physicists, chemists, and test pilots are essential; but no less so are economists, businessmen, diplomats, and a host of others.

No one with even the most primitive knowledge of the subject can possibly believe that any dozen or so men could build and operate a functional space ship, or, for that matter, survive interstellar isolation for the required period and return to their home planet....[10]

In this he was certainly proved right, as within the decade, the effort to get a "man into space" took legions of all the above mentioned skill sets in both the US and USSR. The Apollo lunar landing program increased that need tenfold.

By now von Braun has helped us to understand that a trek to Mars will be a large endeavor. But just how large is surprising. Here are the bones of his magnificent plan:

The study will deal with a flotilla of ten space vessels manned by not less than 70 men. Each ship of the flotilla will be assembled in a two-hour orbital path around the earth, to which three-stage ferry rockets will deliver all the necessary components . . . once the vessels are assembled, and "in all respects ready for space," they will . . . begin a voyage which will take them out of the earth's field of gravity....[11]

In broad terms, von Braun suggested the construction of a Mars-bound space navy to mount his assault upon the red planet. Given the state of technology at the time, it made sense to go large—vacuum tubes were the height of electronic design, and his Mars ships would have looked like WWII submarines inside, all valves, levers, and analog dials. His ideas about redundancy and numbers of vessels for safety were spot-on, but would have cost a fortune. Even his adopted homeland, the affluent US, could never have afforded such a program.

To build his Mars fleet would require another armada—a fleet of forty-eight Earth-to-orbit shuttles (he called them "ferrys" or ferries) that von Braun envisioned for the delivery and assembly of his ten-ship Mars fleet in orbit. The ferries themselves would have been quite an accomplishment, never mind the Mars "flotilla." Given the expense and difficulty of building a fleet of five NASA space shuttles in the 1970s/1980s, and flying them on a far, far less ambitious schedule, von Braun would likely have been shocked at the challenges that his program would encounter—certainly Congress would have been.

In one version of the plan, von Braun's ferries would launch from Johnston Atoll in the central Pacific, which since WWII had been a parcel of extended American territory. A total of 950 launches would have occurred from the island over the course of eight months. That's a lot of launches, fuel, supplies, and logistics to be conducted from anywhere, much less a flea-speck island far out in the mid-Pacific.

An alternative suggested location was on US Air Force property in Cocoa, Florida. This is, of course, generally where the Cape Canaveral launch facilities are located, and it would have been the right choice in logistical terms.[12] The geographic preference, however, would have been Johnston, as it's closer to the equator (which requires less fuel to reach the same orbit), and its remote nature could have been an advantage. But it would have been a difficult program to implement.

Johnston Atoll's total area is about 600 acres, about one and a half times the size of the UCLA campus, comprised of flat, sun-blasted sand and coral—lots of coral. It was enlarged by dredge and crushed coral as the US military sought to make it useful after WWII, and was used alternately as a refueling base, an airstrip, a nuclear and biological weapons testing facility, a missile base, and a chemical weapons disposal facility. It is 3,300 miles from the US West Coast, and is only sixteen degrees north. So, as noted, this would require less energy for the nearly one thousand launches that would depart from there at a rate of 100 per month, or about 3.2 launches per day, seven days a week.

The orbital ferries would be 190 feet tall and about sixty-five feet wide at the base, with wings on the third and piloted stage to facilitate a return to Earth and reuse. The unpiloted first and second stages would parachute back to land in the ocean, with their downward velocity slowed by braking rockets at the last moment. They would then be fished out of the sea and refurbished. This presages the recovery design of the space shuttle's solid rocket boosters (SRBs) that were retrieved in a similar fashion, towed back to Florida by tugboats, and reused. It's noteworthy that each of the SRBs required extensive (and expensive) refurbishment before another flight, and they were, in essence, relatively simple, large, steel-cased skyrockets with solid fuel for propellant. Von Braun's earlier designs would have been *much* more difficult to recondition after their saltwater dunking.

The spacecraft would have a total mass of about 7,000 tons when fueled. When you realize that von Braun's mammoth Saturn V moon rocket measured about 363 feet tall and thirty-three feet at the base, and had a mass of about 3,100 tons when fueled, you understand his orbital ferry to be a truly gigantic spacecraft . . . one that would launch thrice daily, if the weather held up and the atoll wasn't blown to smithereens by a misfire or fuel-handling mishap.

The first stage would have fifty-one rocket engines (compared to the Saturn V's five), with a combined thrust of about 28 million pounds (the Saturn V, the most powerful rocket in history, created 7.5 million pounds). The second stage would have thirty-four engines (Saturn V = 5), and the third stage five (Saturn V = 1).[13] Seeing that it took about five years to get the Saturn V built and tested, you can imagine the challenges they would have experienced getting the monster ferries built, and made reliable enough for 950 launches.[14]

While you're wrapping your mind around the audacity of this plan, just the first phase of the mission to Mars, note that the fuel of choice for the ferries was to be hydrazine and nitric acid. These (especially hydrazine) are amazingly dangerous and caustic chemicals. Both have been used since the early space age in rockets—operating from large, well-protected launch facilities—with much success, but handling them is tricky business. You already know what acid is like, and nitric acid, while far from the strongest form of the bunch, is difficult to handle in large quantities. Hydrazine, the other component, is incredibly reactive and dangerous. So handling the vast volumes of these powerful chemicals on a tiny island would have been a challenge in itself. Over 5,300 tons of fuel would be used per launch, or almost 17,000 tons per day. That's 5,300,000 tons of fuel for the program.

As the ferry rockets thundered out of the mid-Pacific, the Mars fleet would slowly take shape high overhead. The armada of ten ships would include seven ships for the crew and the supplies needed for the transit to Mars and back, plus three one-way cargo vessels, all flying in formation. Von Braun's designs for the interplanetary craft were ingenious, looking almost skeletal when compared to the orbital ferries, since he realized that in the vacuum of space much less would be needed in terms of structure and streamlining. One of the reasons he chose the aforementioned toxic chemicals for fuels was due to the fact that they did not need to be kept critically cold, unlike the far more powerful liquid hydrogen/liquid oxygen combination. There were concerns about handling liquid hydrogen and liquid oxygen, since these are "cryogenic" fuels, meaning that to be maintained as a storable liquid, they would need to be kept very cold—the liquid hydrogen at −423°F and the liquid oxygen at a more balmy −297°F. This has since been accomplished on a number of rocket designs, but to date the longest storage period of cryogenics in space was during the Apollo flights, which took under thirteen days.[15] With a round-trip cycle to Mars of years, and the lack of any practical experience in handling these quantities of extremely low-temperature fuels, von Braun's concerns were certainly warranted.

And there was another factor. Cryogenic propellants would require large, insulated metal tanks. Though hydrazine and nitric acid were touchy chemicals on Earth, they would not be difficult to store in space. His design for the interplanetary ships utilized flexible tanks made of synthetic materials. These would be empty and collapsed for launch, with the fuel shipped up on a later ferry flight. The tanks would then unfold and extend when filled in space. This would require docking and fuel transfers in orbit, another unknown, but something that he thought was doable. It made sense as it was laid out. There were four main thrusting maneuvers for the Mars-bound flotilla—leaving Earth orbit, braking into Martian orbit, leaving Mars, and braking upon return to Earth. Each of these phases would utilize a separate set of fuel tanks—big, nylon-reinforced bags—which would be tossed overboard upon being emptied to further reduce mass. It was like tossing gas cans out the window as you drove cross-country in a pickup truck customized to use multiple fuel containers—each time you tossed one of the cans, you would get rid of some weight and become more fuel efficient—except on a much larger scale.

The Mars ships would look more like underwater research craft than the kind of rockets we are familiar with—a trusswork with spherical fuel tanks and cylindrical sections for cargo or crew. The passenger ships, as he called the seven that would carry the crew, would also have a spherical section at the front for habitation quarters—ten men to each of the seven passenger vessels. There the crew would reside, either strapped into reclining metal chairs or with magnetic boots clanking around on a metal gridwork floor, nicely warmed by all the heat-generating vacuum-tube electronics necessary for the primitive computers, radios, and other necessary equipment. Von Braun proposed that the passenger and cargo vessels, when filled with fuel, people, and supplies, be of the same exact mass to make the collective maneuvers for station-keeping, or formation flying, simpler on the voyage to Mars. For the voyage home, mass would be less of a concern, since the cargo vessels, and various engine stages for the passenger vessels, would be left behind.

Once the fleet arrived at Mars and entered orbit, the crew would survey and explore the surface via telescopes before departing for landing. Three landing craft, which he called "landing boats," would be used for getting to, and (hopefully) returning from, the Martian surface.

The landers were large torpedo-shaped rockets with a wide wing-span, wide enough (he thought) to create sufficient lift in the thin Martian atmosphere—which was, in reality, far, far thinner than he had accounted for.[16] What a surprise his glider pilots would have encountered.

The first lander would set down in a polar region on skis. It was pre-sumed that the Martian surface would be too rough to land in the sub-polar (non-icy) regions, so a slick ice field was to be picked out from orbit. This lander's cargo capability would be much larger than the remaining two, because it was intended to be left on Mars and carried no fuel load for ascent. Its sole purpose would be to deliver a small crew and a pair of tracked expeditionary vehicles, with two cargo and crew units coupled to a tractor. Once these vehicles were unloaded and assembled, the crew would make an epic overland trek to the equatorial regions to build landing strips for the other two "landing boats" that would set down with wheels, like airplanes. This traverse was estimated to be over 4,000 miles. Von Braun was not shy in his ambitions or expectations.

Including this overland journey and the research and exploration time, and readying the remaining two landers for a return to orbit, the landing party would spend an estimated 443 days on the Martian surface. Again, at the time this seemed reasonable—von Braun was operating largely off data from Antarctic expeditions and the like. Knowing what we do today about the true nature of Mars, this kind of mission—even from a perspective of simple endurance—would have been impossible with that kind of technology.

When the surface mission was completed, the crew still had one major task to perform before departing: using a crane on a tractor, they would disconnect the wings from the two "landing boats," then hoist them into a vertical orientation. The formerly horizontal landers would now be vertical ascent craft, looking much like scaled-up V-2 rockets. They would ascend from the surface using the extra fuel carried down from orbit (which was the reason for putting much of the mass needed for surface operations on the first polar lander).

The ascent craft would rendezvous with the orbiting fleet and its skeleton crew, then prepare to depart Martian orbit. Of the ten ships that would arrive at Mars, only the seven passenger ships would return to Earth—the three cargo vessels would not be needed.

Upon reaching Earth orbit, the seven interplanetary vessels would rendezvous with the space station, transfer the crew to orbital shuttles, and home they would go—after almost three years in space.

There were a number of areas that von Braun admitted in the book he had not yet studied in depth. These included navigation, communications, the density of possible meteoritic materials adrift in space, a full development cycle of the spacecraft, and the ability of humans to survive the long-duration space voyage in zero gravity. Virtually unknown at the time, the radiation emitted from the sun and coming in from outside the solar system is far more powerful than expected; the exposure of the crew for that length of time would have been injurious or deadly. If the project had come together and flown before the advent of robotic probes that indicated severe interplanetary radiation levels, there could have been a number of unpleasant surprises in store for the crew.

The Mars Project was a serious and disciplined attempt to design a realistic plan to explore Mars, as approached from an engineering perspective. Von Braun's substantial gifts in political maneuvering would only emerge later as he ran the Marshall Spaceflight Center for NASA in the 1960s. But his Mars mission design did have an impact. Besides convincing readers of *The Mars Project* that such an undertaking was possible, it also attracted a major magazine to do a splashy series of articles based on von Braun's ideas, and brought him to the attention of Walt Disney, who engaged von Braun to work with his studio on a series of extremely popular TV shows and educational films about spaceflight (see chapter 6).

Looking back on this plan from today's perspective it seems almost quaintly naive—given what it took to get three men to the moon on each Apollo flight, his notions for a maritime-styled space fleet and the 4,000-mile overland trek on that hostile planet would be impossible. Those are physical and political realities. But given what was known (or suspected) at the time, it was a reasonable design for a deep-space trek to Mars. And in the long haul, it got a lot of people excited about the idea of exploring that planet, especially after the media outreach that hinged directly and indirectly from the book. Von Braun's first great mission of space exploration took place only in his mind and in the pages of *The Mars Project* ... but set him on the path to become a critical part of Project Apollo's lunar landing effort.

Fig. 3.3. Wernher von Braun (right) poses with Walt Disney (left) in 1954 during their partnership to popularize von Braun's ideas about space travel— including an orbiting station, the moon, and of course Mars, on the popular television show *Disneyland*. Between the *Collier's* magazine articles and various appearances on Disney's television shows and educational films, von Braun's image as a creator of Nazi wonder weapons was slowly cleansed. Image from NASA.

In an interesting twenty-first-century parallel, a long-range sea-ice trek was conducted by NASA planetary scientist Pascal Lee and his team between 2009–2011 that was somewhat analogous to von Braun's plans for a great expedition from the Martian polar regions to its equator. The trek was called the Northwest Passage Drive Expedition.

In the late 1990s, Lee, then a postdoctoral scientist at NASA Ames Research Center, established a Mars analog base (a research station dedicated to conducting Mars simulations) on Devon Island in the high Arctic. The island is home to Haughton Crater, a large and ancient meteorite impact crater. Lee had proposed the site as a good location to simulate Mars fieldwork on Earth, and initiated the Haughton-Mars Project (HMP), an annual field research program focused on Mars analog studies.

In 2003, the HMP began operating a specially outfitted Humvee, donated by AM General, to serve as a Mars pressurized rover simu-lator called Mars-1. But after a few summer seasons in the punishing terrain of Devon, it was clear that a second vehicle was needed to carry out safer, dual-rover traverses. AM General donated a second Humvee, dubbed the Okarian, after a race of Martians in the Edgar Rice Bur-roughs novels about the red planet.[17] But to see service, they would have to drive it to the HMP base.

In 2009, Lee and a small team of fellow adventurers left the coast of the North American mainland, near the westernmost part of Cana-da's Nunavut territory, to drive across nearly 1,000 miles of sea ice and rough terrain to Devon Island, along the fabled Northwest Passage. It took two years—at about the halfway point, Lee had to leave the Okarian to await fresh winter ice—but after a long and treacherous journey, a tired but triumphant team arrived at the Haughton Mars Project base on July 20, 2011 (the forty-second anniversary of Apollo 11's lunar landing).

While the traverse was just a fraction of the 4,000-plus-mile journey von Braun's Mars explorers would have faced, and Lee's team was not burdened with pressure suits or a vacuous atmosphere, the trek was still instructive, and there were parallels to *The Mars Project*'s projected expedition.

"Von Braun's planned polar journey was in my thoughts during our long, cold drive across the arctic sea ice; it was a bit nostalgic in that way," said Lee five years later, during the premiere screening of the feature-length documentary film *Passage to Mars* that was made about the expedition. "While the Northwest Passage Drive Expedition was primarily a logistical trek and we were focused on getting the Okarian to Devon Island, not in doing a high-fidelity Mars traverse simulation, our ice journey was still reminiscent of past Mars exploration scenarios. It did enact, in a way, one of von Braun's early visions about a Mars overland expedition."[18]

CHAPTER 4

PROJECT ORION:
WE COME IN PEACE
(WITH NUCLEAR BOMBS!)

CLASSIFIED: *DECLASSIFIED IN 1979*

It could have been just like the movies. Specifically, the soppy sci-fi melodramas of the 1950s, those humorless, grim-faced sagas of men (always white Americans), square-jawed and broad of shoulder, who faced that Great Unknown, *outer space* (cue the reverb) with stoicism and Yankee guts. The troupe of six to twelve individuals were usually clad in faded blue jumpsuits (probably because they were all of military bent, possibly US Air Force)—no space suits or helmets for these guys; worrying about decompression is for sissies. These were steely-eyed, anvil-chinned rocket men. The heroes would walk up a ramp or climb a ladder into the great, gleaming, cigar-shaped silver *rocketship* (a long-lost term widely used in the early 1950s) without assistance or fanfare—in that sunny postwar era, it took only a handful of servicemen and a few elderly scientists to launch a manned rocket. Once inside, the crewmen would close a submarine-style hatch, strap themselves into great steel chairs, take one last look around their girder-festooned, capacious cabin (1950s rocketship flight decks were the size of your average New York bachelor pad and built like battleships), nod silently to the eldest of the bunch (usually wearing colonel's eagles), who would then push *the button*. This was inevitably a large red push button, marked in true military parlance with something like "IGNITE ROCKETS" or more simply "FIRE!" and off they would go into the Wild Blue Yonder, while on the ground (in a similarly military posture, perhaps within a Quonset hut in New Mexico), a few worried guys in white lab coats watched a twelve-inch radar screen with a huge white dot ascending. A handful of servicemen usually stood nearby, looking vacuously at meaningless blinking lights dancing on their consoles. A single computer, the size of a small RV, would click and whir nearby. This was *Space Command* (or some other imagined, militarized NASA precursor) after all.

Upon reaching space, the colonel would grasp an ice cream cone–sized microphone cabled to the control panel, and as he looked in awe at a receding Earth on the giant "televisor" screen, he'd announce in dour tones, "This is spaceship X-1. We are in outer space." It was all very dramatic and thematically colorless. If you don't believe me, check out the classic 1950s cinematic space extravaganzas *The Conquest of Space* or *Destination Moon*, staples of the genre. Be sure to watch closely during the launch scenes, as the actors' faces are distorted by the horrifying, and as yet little understood, g-forces of launch. Within moments the 737-sized, single-stage craft was in space—no dawdling in orbit—heading in a straight line for the moon or Mars. It's all very humbling and fun, in a deadly serious fashion.[1]

To be fair to the pioneering producers of these epic motion picture dramas, little was known of spaceflight before the 1960s, and sci-fi movie budgets were puny. Few movie studios took the genre seriously, and it's amazing that these innovative moviemakers pulled off what they did, given the general lack of respect these drive-in, Saturday matinee potboilers gained for them.[2] But as we now know, the dramatic scenario outlined above is not exactly how human spaceflight turned out.

But it could have been.

The Apollo lunar landing program, initiated shortly after these types of films were made, mandated a different approach. NASA's moon rocket, Wernher von Braun's masterpiece, would be a multi-stage affair, operating right at the edge of its weight-lifting capability. NASA's first plan was to ascend directly to the moon, land, then, after a suitable period of exploration, return to Earth, shedding stages at appropriate junctures. But this brute-force methodology would have required a truly massive rocket (it was to be called Nova, and was much larger than its successor, the Saturn V), well beyond the means at hand. A bit more planning and a lot of innovative thinking resulted in the moon program we all remember, with the still-massive 363-foot Saturn V rocket propelling a tiny capsule and lander to the moon, of which only the thirteen-foot-wide capsule returned. It took hundreds of thousands of people to build it, thousands to launch and operate it, and somewhere north of twenty billion 1960s dollars to finance it. Apollo was a far cry from the rocketships of the movies.

But there were alternative plans for a massive, battleship-sized single-stage spacecraft that could have flown to the moon and beyond. In its ultimate form, this behemoth would have dwarfed the motion picture versions. A hundred or more crewmen, leaning back in space-age versions of Barcaloungers, would have departed Earth with enough fuel, life support, and supplies to reach the moon, Mars, or even Jupiter and Saturn within months. Once in space the crew would have unbelted themselves and had far more room to drift, eat, work, and sleep than the International Space Station and even most modern submarines offer. It would have been like a well-appointed office complex in space, a true space liner—this majestic craft could have unlocked the entire solar system to exploration within the decade. And best of all? It was *atomic*.

The massive spaceship was called Project Orion (no relation to the modern shuttle-replacing spacecraft beyond the cool name), and it would have been a nuclear-powered behemoth. Orion was first formally conceptualized in a 1955 study by Stanislaw Ulam, a Polish American mathematician who was part of the Manhattan Project in WWII, and Cornelius Everett, working from notions that Ulam had first pondered soon after WWII. Besides working on the bombs dropped on Japan, Ulam was, along with Edward Teller, a prime mover on America's first hydrogen bomb project. Soon after completing his work on H-bombs, Ulam formalized his thoughts about nuclear rocket propulsion. Other work was being done on atomic rockets, but was less dramatic—these projects involved superheating a fuel mass, such as liquid hydrogen, inside a fission reactor to eject it at high speeds out of the rocket nozzle. While much more efficient than the chemical rockets being designed by von Braun and others, it was not the massive leap in propulsion that would take humanity to the stars. Ulam had a different idea—nuclear pulse propulsion, which was not fully declassified until 1979.[3] From the abstract:

> Repeated nuclear explosions outside the body of a projectile are considered as providing means to accelerate such objects to velocities of the order of 106 cm/sec.[4]

Yes, that's right. Rather than fiddling around with rapidly expanding heated gasses with a nuclear reactor, Ulam took the most direct path to high energy release: nuclear explosions. Ulam had been mulling this over for more than a decade, reasoning that chemical rockets were terribly constrained by both the mass of the fuels and the temperatures at which they could realistically operate. Other proposals to detonate tiny nukes inside combustion chambers (one proposal suggested a chamber diameter of 130 feet, or almost four times the diameter of the Saturn V), while an improvement over chemical rockets, were deemed impractical, and did not offer a large enough increase in performance to impress Ulam. But what if the combustion chamber could be eliminated altogether and a small nuke simply detonated in open space? A percentage of the energy released by a reasonably sized nuclear explosion—not specified in the paper, but probably on the order of a half to one kiloton (about 10 percent that of the Hiroshima bomb)—would nudge a nearby spacecraft with propulsive force that, while brief, would be enormous.

Ulam characterized the spacecraft as an unmanned thirty-three-foot diameter, disk-shaped ship, with a mass of twelve to twenty tons. It would experience an acceleration of up to 10,000 g (the Apollo astronauts, riding atop the Saturn V, maxed out at just under 5 g, though the rocket was capable of more)—hence the unmanned nature of the design. Human occupants would have been turned into puddles of red jelly within moments. This robotic probe would carry dozens to hundreds of bombs, to be released at roughly one-second intervals (accompanied by a disk of plastic or container of water that would vaporize when the nuke ignited, to enhance the effect), and the resulting force of these continual explosions would propel the craft forward—right *now*.

Ulam was concerned about the heat impinging on the base of the craft, and suggested that a magnetic field might help to shield the spacecraft from the high-energy, one-millisecond flashes.

This was about as far as he got—it was a short study, but an intriguing one, and did not go unnoticed. In 1955 a new company called General Atomics was founded. It was a subdivision of General Dynamics, a huge defense contractor and builder of military submarines. General Atomics would specialize in efforts to harness the recently liberated power of the atom—in effect, their mission would be to find profit in nondestructive uses of atomic fission. The company became involved in a number of ventures, including a commercial nuclear reactor power generator, which was widely deployed. They also became interested in Ulam's classified paper (to which the chiefs of the company were apparently privy), and decided to pursue a serious study of the completely theoretical ideas within. Thus was born Project Orion, the nuclear pulse spaceship.

Theodore Taylor, who held a PhD in physics from the University of California at Berkeley, had spent eight years at the Los Alamos National Laboratory and was a recent arrival to General Atomics. He was put in charge of Orion, and was partnered with a young Freeman Dyson, who held a PhD from Cornell. Between them, they made a thorough study of the propulsion concept, with variants ultimately ranging from a "small" 10,000-ton craft to an interstellar-capable version that would have been an unimaginably massive eight-million-ton hulk.[5] While this seems to be a huge and possibly unrealistic range (a bit like building a Death Star that can fly into space from the Earth's surface), it does go to show the flexibility of the propulsive concept. Nuclear pulse propulsion scales nicely, if you have the engineering and technology to back up your ideas.

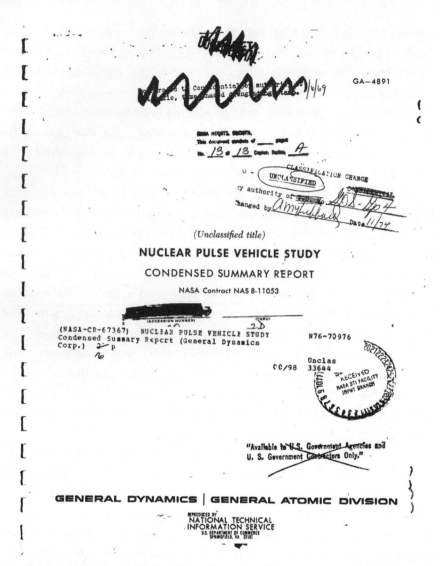

GA-4891

(Unclassified title)

NUCLEAR PULSE VEHICLE STUDY

CONDENSED SUMMARY REPORT

NASA Contract NAS 8-11053

(NASA-CR-67367)   NUCLEAR PULSE VEHICLE STUDY
Condensed Summary Report (General Dynamics
Corp.)   2 p

N76-70976

Unclas
CC/98   33644

"Available to U.S. Government Agencies and
U. S. Government Contractors Only."

GENERAL DYNAMICS | GENERAL ATOMIC DIVISION

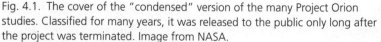

Fig. 4.1. The cover of the "condensed" version of the many Project Orion studies. Classified for many years, it was released to the public only long after the project was terminated. Image from NASA.

Taylor ran Project Orion to maximize output on his small budget—he let his people do what they did best, and gave them latitude to accomplish their tasks generally how they desired. Dyson at one point said that he thought Taylor had taken inspiration from the VfR rocket society that von Braun had been engaged with before WWII, which is entirely possible—within the limits of US Army oversight, von Braun was trying to do something similar in Huntsville, Alabama, at the same time, and getting excellent results.[6]

This project became increasingly expensive, driving General Atomics to approach the Advanced Research Projects Agency, or ARPA (DARPA's predecessor) for additional funding. ARPA responded with a study budget of $1,000,000 per year to underwrite the project.[7] It doesn't seem like a lot today, but Project Orion was being fueled primarily by brainpower at this point, and with the average annual salary in the US in 1958 being about $3,700.00,[8] a million dollars went a long way toward salaries and talent retention.

Taylor and his crew refined and improved Ulam's designs. The bomb sizes were determined for various versions of the spacecraft, and additional reaction mass, now defined as either plastic or wax, would probably be bonded to the bombs. The resulting craft looked either like a giant beehive or perhaps like a huge .50-caliber bullet standing on a milk stool.

Perhaps the most important change to Ulam's design, however, was that the General Atomics version would be *manned*. The top half contained crew quarters and supplies storage; the bottom half was fuel (bomb) storage and a shock-absorbing system—*really* big, heavy-duty shock absorbers—capable of converting the 10,000 g, crushing propulsive blasts into survivable acceleration by absorbing and gradually releasing all that energy. At the base was the "pusher plate," the wide, flat disk that would absorb and transmit part of the energy from the bomb blasts. The bombs were to be sent from the storage magazines through a tube that penetrated the pusher plate, to detonate beyond—close enough for the blast to be effective, but far enough that the short-duration pulse would not melt the plate, a distance of about 100–200 feet. The bombs were uprated to twenty kilotons, in the same general range as the Hiroshima bomb, and would detonate about every ten seconds.

The manned version of the spacecraft being studied in the late 1950s was to be about 150 feet high and 135 feet in diameter at the base (the pusher plate). Mass at liftoff would have been 10,000 tons (The Saturn V was about 3,100 tons). But rather than burning most of its mass (as fuel) to get into space, as the Saturn V did (and any chemical rocket does), most of Orion's mass would actually end up *in* space—a huge, and very helpful, difference. With a capacity of 2,000 bombs in this design, Orion's reach would be vast. As Dyson, ever eloquent, put it, "Mars by 1965, Saturn by 1970."[9]

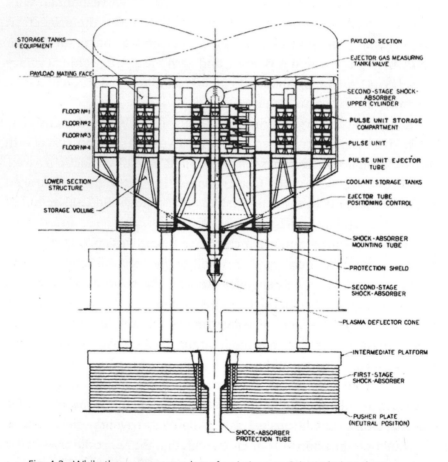

Fig. 4.2. While there were a number of variations on Orion's design, the propulsion system remained the same. At top is the "magazine" for holding the atomic bombs, from hundreds to thousands of them depending on the version. The tubes extending downward are shock absorbers to mitigate the massive concussions on the pressure plate, at bottom, from the repeated nuclear explosions. The center tube, leading to a hole through the pressure plate, feeds the bombs to the rear of the spacecraft for detonation. Image from NASA.

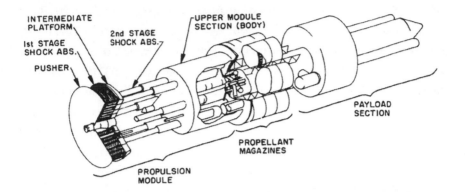

Fig. 4.3. This later configuration of the Orion design is downsized for launch on conventional large rocket boosters such as the Saturn V, removing the risk of radioactive contamination by the use of atomic bombs to achieve Earth orbit. Image from NASA.

This was all being planned at about the same time America was struggling to get a single man into space for three orbits, jammed in a tiny Mercury capsule, set atop a chemically fueled Atlas rocket (kerosene and liquid oxygen). Orion, in stark and awesome contrast, would carry a crew of 150 men, with thousands of tons (not pounds, *tons*) of supplies. They could go just about anywhere they wanted, assuming that life support and other issues could be worked out. No wonder the General Atomics researchers felt that von Braun was on the wrong track—Chemical rockets would never open up the solar system for spaceflight the way atomic pulse spacecraft like Orion could.

Orion would require a continuing stream of nuclear explosions to accelerate. The frequency of detonations varied depending on the version being studied (its mass and intended purposes) and the phase of its flight. Once in space, the explosions could be spaced farther apart depending on the desired acceleration. Also, the bomb yield varied according to the flight phase—they could carry a variety of bomb sizes for different needs. For example, for liftoff they would have to detonate smaller bombs more frequently—at least one 0.1 kiloton bomb per second. And yes, they were going to do exactly what you are probably thinking—they would launch their atomic monster from Earth. Bang, bang, bang. Sorry, you have cancer—we're on our way to Saturn.

This presented a thorny problem. To avoid the wrath of angry citizens and (later) the Environmental Protection Agency (EPA), the designers would either have to reduce Orion's mass to allow it to be lofted by a single launch of a Saturn V (which was later adopted into a revised Orion plan), or launch it in bits and pieces on multiple Saturn Vs and assemble it in orbit, or just go ahead and launch it under its own power with the nuclear bombs if the government could be persuaded to agree. Dyson calculated that each launch, statistically speaking, could result in between one and ten deaths due to cancer induced by fallout.[10] Of course, compared to car accidents, cigarettes, or heart attacks, that number is small. But car crashes, cancer, and heart attacks are accidental—launching a giant atomic rocket that spews radioactive waste into the atmosphere is deliberate. It was bound to run afoul of popular opinion.

In 1959, ARPA decided that it was done with Project Orion. NASA had taken over manned spaceflight projects (with the exceptions of some stillborn US Air Force efforts) and was, at the time, not interested in nuclear pulse propulsion—their rockets used explosive chemical reactions to go into space. They did use nuclear materials to power lunar experiments on Apollo, and some of their robotic planetary missions, but in small and nonfissionable quantities. So Taylor and General Atomics approached the US Air Force. Their timing was excellent, because it coincided with the air force's efforts to take a piece of the space pie for itself. They wanted to launch their X-20 spaceplane into orbit, or maybe beat NASA to the moon, or perhaps build a moon base, or . . . well, you will know that story by the end of the book. The air force had a wide-ranging agenda that never amounted to much in terms of successful manned spaceflight.[11] And, to put it kindly, they viewed nuclear explosives in a somewhat different light as did Taylor and his cadre—which is to say that nukes were useful for the annihilation of Moscow, but harder to envision as a propulsion source. Nevertheless, they eventually agreed to provide funds for the project with the hope that it might eventually have a military application.

The Orion adherents began to study a design for the spacecraft that might be able to integrate with NASA's Saturn V and their overall plans. It would be smaller and far less ambitious than their original ideas, scaled down to a bare minimum, but it would be a way to salvage the project. To understand just how far downscaled it would be, we must look at the propulsive force availed by nuclear pulse propulsion.

Rocket thrust can be measured in a few ways, and one of them is specific impulse, or ISP. It measures efficiency per unit of fuel consumed by the rocket, and can be indicated in seconds.[12] So the Saturn V, while having a thrust of 7.5 million pounds,[13] had an ISP of about 263 seconds at sea level. The shuttle, using cryogenic hydrogen and oxygen, was about 450. The atomic reactor rockets described earlier—that heated a liquid fuel with heat from the reactor—were about 900. The atomic-bombs-in-a-combustion-chamber design was about 1,150. But the original designs for Orion ranged from—hold on to your hats—10,000 to *one million* seconds, or up to 3,800 times that of the Saturn V.

But without the massive thrust from a nuclear-powered launch, and limited to the carrying capacity of the Saturn V (which used relatively low-energy kerosene and liquid oxygen for power), this Orion variant would have to be scaled down to something that could be carried by NASA's biggest rocket. They returned to the idea of a vehicle of about thirty-three feet diameter (the same as the Saturn V's first and second stages), weighing about 100 tons. The ISP of this design was down—way down—to about 1,800–2,500 seconds, shameful by nuclear pulse standards, but still at least seven times better than the Saturn rocket. In a business where grams matter (and cost a fortune to launch), that's a hell of a deal.

While the "Orion Mk. II" was sized to fit on the Saturn V, it could have taken up to four launches to get even this smaller version into space. It was still unclear how many Saturn V rockets would ultimately be manufactured, and they were expensive. But the advantages offered by even a smaller Orion, once in space, would have been vast. NASA was just getting started on the Apollo program, but already had

Mars in its sights. Using just a Saturn V–class rocket, or even a number of them, a chemically fueled Mars rocket would take at least five to seven months one way to the red planet. The small Orion would take less than four months. The Saturn V–powered mission could carry a small crew with limited cargo and not much space; small Orion could carry eight crew members and 100 tons of cargo. Even von Braun, who believed in conservative engineering and chemical rockets, became an ardent supporter of Orion (he was, somewhat uncharacteristically for people in a position of technocratic power, usually willing to accept an outside idea if the math checked out). But Orion had no future, and this study would be its last hurrah for decades.

In 1963 the US, USSR, and UK signed the Limited Test Ban Treaty. This was designed primarily to slow weapons development and testing on Earth and in the atmosphere, but also applied to the oceans and "outer space." The last bit was the final nail in Orion's coffin. Nuclear explosions in the atmosphere were now even more of a nonstarter.

Dyson had since departed the project, and had in fact been instrumental in the treaty (he had nothing against Orion, though he did have second thoughts about the cancer risks). But Taylor and others still believed in its value, and pushed for a few more months. The end came in 1964; the air force was unwilling to continue without help from NASA, and NASA was already stressed to meet the demands of the Apollo program. Orion was cut adrift.

It was a premature end to a program that could have returned great results. As Dyson put it, "This is the first time in modern history that a major expansion of human technology has been suppressed for political reasons."[14] While there were legitimate concerns about the use of atomic explosives for launch, and even the carrying of nuclear materials into space for use only outside the Earth's atmosphere, the reasons for the cancellation were primarily political and image-driven. NASA was perceived as a "clean" agency, and the small amounts of nuclear material it did fly was done so with relatively little fanfare.[15] US and Soviet military flights carrying nuclear reactors aboard were even quieter.

There would have been many other challenges to bring Orion from paper to reality. Supporting a large crew would have required vast life support systems that were not worked out until large submarines—ironically also powered by the atom—began routinely prowling the world's oceans in the 1960s. Also, as discussed in chapter 3, the intense radiation encountered in interplanetary spaceflight was not well understood until later, and Orion would have required extensive shielding and protective "vaults" for the crew. The list goes on.

But with its incredible propulsive power, Orion would have been able to carry the equipment available in the day, without the expensive miniaturization that was required to make the Apollo missions to the moon possible. Large stocks of oxygen, food, and water could have been carried aloft, reducing the need for modern, regenerative life support systems. Radiation shielding could have been designed using traditional materials—the large water supply would be one barrier, and even lead, the traditional solution, might have been used. The crew would have been traveling inside a veritable battleship in space, with all the attendant comforts of home, and the solar system and its mysteries could have, in theory at least, been explored before the end of the twentieth century. The basic engineering was and, according to some who continue to study such spacecraft, is *still* sound. All that's needed are a few hundred spare nuclear bombs and the nerve to use them.[16]

CHAPTER 5

LUNEX:
EARTH IN THE CROSSHAIRS

CLASSIFIED: *DECLASSIFIED IN 2014*

Not content to let the army claim all the cosmic glory that was surely just around the corner in the early space age via programs like Project Horizon (see chapter 2), the US Air Force attempted to muscle its way into manned spaceflight with its own lunar program, LUNEX, delivering a proposal to the federal government in 1961. And this was not just a suggestion—to put it in the words of the air force blue-suiters, "Normally the end product of this type of study is an Evaluation Report. However, due to the importance of the study conclusions and the significance of time, it was decided to prepare a Program Plan, as part of the final Report."[1] In other words, get off your civilian backsides and pay attention—this getting to the moon stuff is goddamned important. . . . The Russians are just over your shoulders. This is military business, and we're ready to go.

LUNEX was not dissimilar from the army's Project Horizon, itself submitted and passed over by the Eisenhower administration just two years before. While less ambitious, LUNEX followed the division of military priorities just as you'd expect. Where the army's Horizon was about a large forward base, capturing the highest of the high ground and holding it against overland incursion by space suit–clad Communist moon soldiers, the air force thought in streamlined and strategic terms, and it did not want to take the time to build the enormous logistical base that Horizon would have required. "Development items should be started *immediately* if maximum military advantage is to be derived from a lunar program," the 1961 proposal read, noting that lunar voyages could be initiated as early as 1967 (italics in the original document).

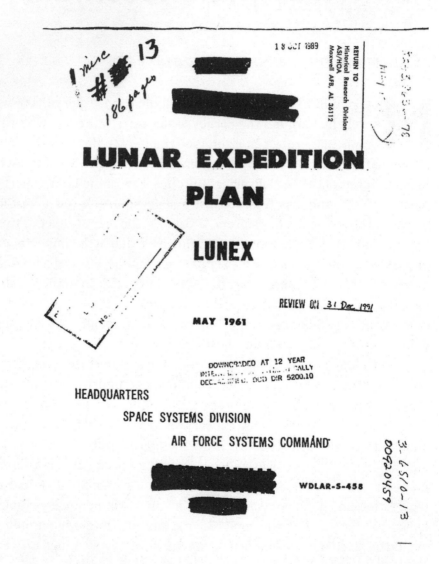

Fig. 5.1. The final report on LUNEX as submitted by the US Air Force. It was classified for years after the project was rejected for funding. Image from DOD/ USAF.

And what of the strategic value? Again, they were militarily blunt:

The final decision concerning the types of strategic systems to be placed on the moon (such as a Lunar Based Earth Bombardment system)[2] can be safely deferred for three to four years. However, the program to establish a lunar base must not be delayed and the initial base design must meet *military* requirements. For example, the base should be designed as a permanent installation, it should be underground, it should strive to be completely self-supporting, and it should provide suitable accommodations to support extended tours of duty.

Of note is "extended tours" and "bombardment system." While little is made of the maintenance of a fort on the moon, LUNEX would clearly give the Strategic Air Command guys, who always had one hand hovering near the nuclear launch button, the ultimate platform from which to dispatch their USSR vaporization package. Not that, in a detailed analysis, it would be superior to an orbiting platform or even ICBMs (it was essentially the same argument—with the same weaknesses—that Horizon had used), but in the late 1950s, with a single computer taking up a large room and not being very capable, human-operated systems seemed the way to go. At the time, conventional knowledge dictated that nukes aimed at Moscow would, in time of war, be delivered on long-range bombers, which could be shot down by the enemy. So high-speed incoming weapons, safely stored 240,000 miles away on the moon until needed, were appealing, despite the travel time. That is, if they survived the flight back to Earth. Spacecraft returning from the moon would have to enter the Earth's atmosphere at about 25,000 mph—a lot of work would have to be done to prevent damaging the warhead. But having a first-strike proof lunar base of operations seemed like an excellent deterrent.

In broad strokes LUNEX was similar to Horizon: get some soldiers to the moon to establish a foothold, and build a military infrastructure there. But the character of the air force bled through the document. Von Braun and the army had taken a "bigger is better" approach to getting people and supplies back and forth to the lunar base: *lots* of big rockets. Not so with LUNEX. It was a leaner, and presumably more achievable, plan.

Central to the effort would be the manned Earth Return Vehicle, or ERV. It would have looked like a cross between the space shuttle and another concept the air force explored in the 1960s, a lifting body. The 135,000-pound ERV was roughly triangular in shape with small vertical stabilizers and stubby winglets at each trailing edge. Its crew of three men would fly in comparative comfort, not wedged shoulder to shoulder the way NASA's astronauts stuck in capsules would be. The ERV was intended to handle reentry speeds up to Mach 35, and was thought to have been reusable if properly engineered.[3]

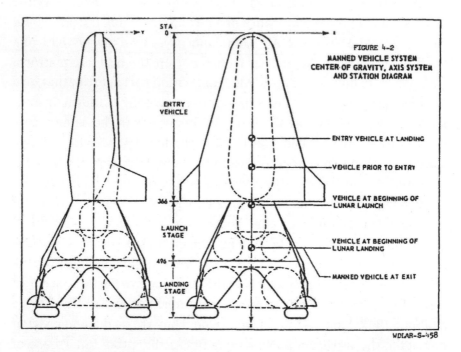

Fig. 5.2. Schematic view of the US Air Force's LUNEX lander and crew vehicle. This was a variation on NASA's early designs for the Apollo program, in which the same craft that left Earth would land on the moon and return the crew to Earth. NASA later adopted the "lunar orbit rendezvous" approach, with a separate crew capsule and lunar lander, both of which were single-use spacecraft. The air force, of course, wanted a winged vehicle that would be flown by pilots, which was more aligned with their tradition of flight. Image from DOD/USAF.

The booster for the ERV was a large cryogenic rocket, fueled by liquid hydrogen and liquid oxygen, as the Saturn V's upper stages (and later space shuttle) were. These fuels pack more punch in a given volume than the Saturn booster's kerosene/liquid oxygen combination. There was also a heavy-lift cargo variant of the ERV that would have been instrumental for building up a moon base. Mention was also made of using solid fuel boosters for its primary propulsion system— something that von Braun did not favor for man-rated rocket designs, but that did eventually make its way into the shuttle program. In the end, either design approach could have worked.

The LUNEX rocket was to be rated at six million pounds of thrust, but with about 10 percent more lifting capacity than the Saturn V (350,000 pounds vs. 210,000 pounds). Of note, the air force had been an early proponent of large liquid rocket engines (key to delivering nuclear bombs to Russia en masse), and had been the first to champion the huge F-1 rocket engine that would eventually send Apollo to the moon.

For LUNEX, a giant three-stage rocket would lift off from Earth, aimed directly at the moon. The ERV had a fairing at the back with a series of landing legs arranged around the perimeter. As it neared the moon, the craft would assume a rear-down attitude and begin to retrofire, braking as it neared the lunar surface. The ERV would land tail-first and essentially blind, but surely those details would be worked out later. In fairness, NASA too was planning a tail-first landing for its early lunar mission designs, possibly using rearview mirrors or oddly protruding cupolas from which an astronaut would be able to pilot the lander. There was mention in the final report, issued in 1961, of considering an Earth orbit rendezvous (EOR) mission design, in which components of the lunar craft would be assembled in orbit with multiple launches, again roughly parallel with NASA's mission planning (NASA would settle, with some reluctance, on the more economical single-launch lunar orbit rendezvous, in mid-1962).

To give credit where it is due, the plan did identify a few more potential complications than the hastily completed Project Horizon document had. Tail-first landing was one; reentry of the lifting body was mentioned as well. But a program of incremental steps, including robotic reconnoitering of the moon and the placement of landing beacons, was expected to overcome these potential problem areas. The air force felt that the technologies needed were simply extensions of those already well understood, and would not require any major breakthroughs to accomplish.

How shocked, then, the authors of the LUNEX study must have been as they watched the (in comparison to their own optimistic estimates for LUNEX) expensive Apollo program struggle with one technical challenge after another over the course of the next six years.

The schedule was as optimistic as the assumptions about the technology. With a presumed starting date of 1962, air force planners proposed a test flight of the system in 1965, a manned circumlunar flight in 1966, and a landing and return of a crew to the moon in 1967.

In what could be interpreted as a stab in NASA's eye, a section of the final report suggested a "Design Philosophy" that included,

> The Lunar Expedition Plan has been oriented toward the development of a useful capability rather than the accomplishment of a difficult task on a one-time basis. The use of a large booster is favoured for the direct shot approach since studies have shown this to be more reliable, safer and more economical as well as having earlier availability. However, another approach using a smaller booster in conjunction with orbital rendezvous and assembly is also considered.

Besides the occasional (and odd, for a US military report) "Continental" spelling ("favoured"), the suggestion of a "useful capability," i.e. permanent infrastructure, rather than a "one-time basis," i.e. Apollo's blunt-body capsule, was probably intended to show the superiority of the LUNEX proposal over NASA's Apollo system. In the end, they may have had a point—a well-designed, robust, and reusable lifting body could have represented a long-term cost savings, as would a permanent settlement on the moon. However, this is far from assured. The later space shuttle, also intended as a reusable, robust, Earth-to-orbit infrastructure, was anything but cost-effective, and turned out to be as expensive (and less reliable) than the Apollo system it replaced. And had the Saturn V/Apollo spacecraft system been built in larger numbers and used for decades, it could have turned out to be an affordable and reliable approach to permanent access to space. For proof of this, one need only look as far as the long-serving Soyuz rocket and spacecraft design of the Russians, soon to enter its fiftieth year of reliable service. As for the "permanence" of a lunar base ... well, nobody at the time seemed to fully appreciate the challenges of living off-Earth, not the air force, not the army, and not even NASA. And the expense of such an undertaking would have been far higher than was calculated.

One more limitation of the LUNEX design concerned abort capabilities during launch. All of NASA's designs until the shuttle included various emergency escape mechanisms to deal with urgent situations between leaving the pad and separating the capsule from the booster—one of the most dangerous phases of spaceflight. Like the shuttle, there would be a period of time during which the ERV would be susceptible to booster failures during launch, possibly resulting in the loss of the crew. To wit:

> In providing an abort philosophy for the Lunar Program it should be noted that the Lunar Re-entry Vehicle, the Lunar Landing Stage and the Lunar Launching Stage all possess inherent abort capability if utilised properly during an emergency. With sufficient velocity the re-entry vehicle is capable of appreciable manoeuvring and landing control to provide its own recovery system. The Lunar Launching and Lunar Landing Stages possess an appreciable delta-v capability that can be used to alter the payload trajectory to better accomplish recovery of the man. However, in either case the manoeuvres will have to rely on computing techniques to select the best possible abort solution for any specific situation.

Note the "if utilised properly during an emergency" language (there's that oddball Continental spelling again). Similar abort scenarios were designed for shuttle launches, but fortunately most of them were never required. While LUNEX had an advantage in that it rode at the top of the launch vehicle instead of on the side (like the shuttle), it did not have the equivalent of the top-mounted escape rockets that the Mercury and Apollo capsules had, which in the event of a failing booster could ignite instantly, whisking the capsule away from the rocket. The capsule would boost to a higher altitude and then descend via parachutes as it would during a normal reentry (Gemini capsules carried parachute-carrying ejection seats for the same purpose). The LUNEX ERV would have to attempt to ignite its lunar descent stage and boost away from the launch vehicle to attempt a horizontal landing— with a far from assured outcome if any damage was incurred during the emergency. Depending on vehicle condition, the status of the pilot, and the altitude of separation, such a landing could have been a very tricky business.

Further language included, "It is well recognised that maximum reliability is desirable, but also known that reliabilities in excess of 85 to 90% are extremely difficult to achieve with systems as complex as the Lunar Transportation System." This was an acceptable risk quotient for the air force—NASA's goal for Apollo was 99.9 percent reliability; realistically, they settled for a bit less, but in any event it was far in excess of 90 percent.[4]

Another possible danger was the descent to the moon, a dangerous phase in any lunar mission design. LUNEX's approach was simple. In the event of complications during the landing: "Where possible the Lunar Launching Stage will be used to attain a direct or circumlunar trajectory that terminates in an earth return. When this is not possible the Lunar Launching Stage will be used to accomplish the safest possible lunar landing. Recovery of the crew will not be provided in this system and selection of the above alternatives will be accomplished automatically on-board. Crew recovery will be provided by another stand-by Lunar Transport Vehicle." In summary, if there were problems before the bullet-like trajectory sent them careening into the lunar surface, the pilot would try to wave-off and loop the moon (complete a partial orbit) to head back to Earth. If the spacecraft was already too close to the surface to accomplish this, they would do their best to land, and wait there for a second spacecraft to come and rescue the crew (if they survived). Apollo's answer was to first enter lunar orbit, and then, if all systems checked out properly (a nice buffer should something go wrong with the lunar module en route), head to the moon's surface. If the descent engine failed during the landing phase, the LM's lower stage could be ejected and the ascent engine (a completely separate unit in NASA's design) would be fired, bringing the two astronauts back to the waiting command module, still safely orbiting the moon. The Apollo lunar flights never had a backup rocket on the pad in case the crew was stranded on the moon, in lunar orbit, or Earth orbit . . . nor was one ever needed.[5]

Electrical power for the ERV was to be supplied using either fuel cells (as Apollo did), or solar panels (as the Soyuz did). Guidance while in spaceflight would utilize star-spotting optics or trackers, as both Apollo and the robotic programs of the 1960s did, assisted by computers onboard and on the ground. Again, the design follows the same general approach as NASA's manned mission planning of the era. Many other factors, such as radiation, dangers from drifting meteoroids, lunar surface operations, and more were listed simply as current unknowns, and were items for further investigation—a very candid approach to challenges that could not be accurately weighed at the time of the proposal, but also an area of enormous potential for budget increases.

A final section of the LUNEX proposal attempted to peer into the cloudy crystal ball of military intelligence and come up with some estimation of what the Soviet Union might be planning in spaceflight. It was clear at the time, in 1961, that spaceflight was a priority for the Russians and they had a clear lead. Their rocket engines were more powerful than ours by a wide margin, and their orbiting spacecraft were larger and heavier. More difficult to assess were Soviet military aims in space, their intentions toward orbiting platforms (possibly military), and, of course, a lunar presence. Sending a human into orbit and bringing him back was one thing; sending a crew off to the moon, landing on its surface, operating there, and bringing them back to Earth was quite another. The technologies required for all the incremental phases of such a flight, which included critical items like life support and deep-space navigation and control, were unknowns, especially when looking at Russian technical acumen. But one thing seemed clear to the intelligence agencies of the West: the Russians were armed for bear. From the final pages of LUNEX:

> The Soviets do not differentiate between military and non-military space systems. They have talked of a peaceful intent of their space program but there are many pounds of payload in their satellites which cannot be accounted for on the basis of data given out. It should be presumed that this could be military payloads. With this in mind, it can be stated that during the early 1970's it is possible that space weapon systems will be developed as a supplement to earth-based delivery systems. It is also possible that military facilities may have been established on or in orbit around the moon.

The implication was, as with Horizon: we had better get there first, and the US Air Force is your best bet, Mr. President.

In the end, LUNEX met a fate similar to Horizon: it was a nonstarter. Despite the fact that it would require a relatively meager budget of only $7.5 billion to make the first lunar voyages, compared to Apollo's final price tag of about $20 billion (in late 1960s dollars),[6] and indicated small staffing levels (only in the hundreds) to ramp up to active status (Apollo would ultimately engage almost 400,000 people), and that the air force had flight experience of which NASA could only dream, the project was not funded to move forward.[7] It did ultimately motivate lifting body designs that would be tested a few years later at the air force's base in the high desert of California—a small but important series of experimental craft were tested concurrently with the flights of the X-15, and met with varying degrees of success. These lifting body experiments, and similar Soviet efforts, ultimately provided data of value to the space shuttle program, and later to companies like Sierra

Nevada Aerospace, now building the *Dream Chaser* mini-shuttle that will soon begin routine deliveries to the International Space Station and, if all goes according to plan, both NASA-contracted and publicly available crewed flights into orbit. But the air force will not be instrumental in these efforts, and has since moved on to unmanned mini-shuttles that are flying as this is written.

In a bit of a second-place victory, the air force was asked by NASA for assistance in the design of the complex management system needed to oversee the Apollo program. The military had valuable experience in the management of vast enterprises—the air force was undertaking an enormous buildup of nuclear missile capabilities in those years—and NASA knew that this could make all the difference in its own pursuits. An air force general, Samuel Phillips, was ultimately engaged as the program director for Apollo. He is one of the unsung heroes of the space race, and his military experience, as well as the complex management structure developed for programs like the Minuteman missile, were instrumental in Apollo's success.[8]

But back to 1961—despite the lack of funding for LUNEX, the air force did not simply shrug and call it a day. Already, plans were afoot that would eventually manifest themselves as orbiting platforms and reusable spaceplanes (both will be seen in later chapters). While President Eisenhower's civilian space agency could enjoy its moment in the sun for now, the air force felt that space was surely its rightful domain, if it could just get the right proposal approved and funded. The generals knew that it was just a matter of time. . . .

CHAPTER 6

THE WHEEL: AN INFLATABLE SPACE STATION

UNCLASSIFIED

We've all seen the International Space Station in its $150 billion glory—a multitude of high-tech metallic modules joined by carefully designed junctions and airlocks, surrounded by skeletal trusswork and a vast field of solar panels, and crewed by an average of six astronauts, making a silent passage over the globe every ninety minutes. But this design, carefully worked out with input from the five original partners in the international venture, was the result of decades of study. There were many designs, and other space stations, that preceded it. The Soviet Union's Salyut and America's Skylab were single-module units, the latter a converted rocket stage from the Saturn V. The USSR's later Mir space station was a smaller modular unit, the closest relative to the eventual ISS. But long before, at about the same time that he was designing moon rockets and Mars expeditions, the ever-imaginative Wernher von Braun had conceptualized something vastly different—an orbiting station with a crew of eighty men that would be, in effect, a giant, rotating rubber tire in space—not like a hard-shelled submarine, but more like a tubular balloon.

A station in space was not a new idea, any more than a spacefaring rocket was. The notion of an outpost in orbit had been around for decades—enterprising individuals had written of them, sketched designs, calculated structural loads, and proposed the many benefits of such an orbiting facility, some since the late nineteenth century.

Fig. 6.1. A later NASA variation on von Braun's inflatable space station design, utilizing easier-to-launch straight segments. Image from NASA.

The story of space stations in popular literature begins with the "Brick Moon,"[1] a whimsical invention of Edward Everett Hale, a Boston clergyman who had an unorthodox vision of a way to get closer to his Creator. In this tale, which appeared in serialized form in the *Atlantic Monthly* beginning in 1869, his protagonist builds a 200-foot-diameter sphere out of masonry and shoots it into space with water jets. He wrote the piece as fiction, thank goodness, but it was an interesting early notion about the rigors of relocating to orbit. The structure is supposed to be a navigational beacon, a second moon at night, but launches with people aboard—the old "accidental passenger" trope.

Whimsy aside, you must by now suspect that the earliest people to seriously consider livable structures in space were going to be either German or Russian, or, in this case, both. Konstantin Tsiolkovsky, that "father of Russian rocketry" who, along with Hermann Oberth, so affected the young Wernher von Braun, wrote of space stations and rockets as early as 1895. Tsiolkovsky also postulated a rocket that would carry twenty crew members and convert into an orbital station that would generate artificial gravity via centrifugal force—spinning along its axis. The passengers would subsist on crops grown in a space greenhouse, flooded by the permanent sunlight of space.[2] He also suggested that such a station could be a way station for traveling beyond Earth. His ideas for other rocket designs and their applications were no less visionary.

Following these space-bound footsteps, German Hermann Oberth wrote of even grander visions. In two important books, *By Rocket into Planetary Space* (1923) and *The Paths to Space Travel* (1929), Oberth laid out ideas for space outposts to act as rocket-refueling stations and for Earth observation. At the time the books were written, these were solutions to problems that few had yet pondered, but this would change as Germany drifted toward war in the 1930s. Oberth also discussed plans for an orbiting mirror, later dubbed the "sun gun," which could be useful for lighting large harbors at night, to cite one example. But the giant orbiting parabolic mirror was recognized by the Nazis for its more nefarious potential—a huge, space-based death ray. Early designs indicate an orbiting mirror a few hundred feet in diameter that would concentrate the rays of the sun onto the planet below. Rather than illuminating marinas, the German military saw an orbiting weapon directed onto unsuspecting cities below, resulting in broad swaths of torched devastation via intense solar energy.[3] An illustrated description of Oberth's designs appeared in *Life* magazine in 1945, citing US Army technical experts as its source. The illustrations show a massive orbiting platform crewed by German soldiers, who would survive by breathing air created by pumpkins growing in a greenhouse (and, one would imagine, later baked into pies via smaller versions of the terror weapon they were staffing). Drawings from the article show an Aryan trooper, dressed in what could be lederhosen, short sleeves, and a small leather cap, tending the crop.[4]

Closer to our tale, in 1928 Herman Potocnik (who wrote under the pen name of Hermann Noordung), an engineer and officer in the Austrian Imperial Army, published *The Problem of Space Travel*,[5] which looked at space stations in some detail. This presented the first published wheel-shaped configuration for an orbiting outpost, and influenced space station design ideas for decades thereafter. The 100-foot diameter station looked like a bicycle wheel, with an airlock at the central hub and two tunnels connecting it to the perimeter of the habitat ring. The hub included a solar power generator that would use steam boilers to generate electricity, and an astronomical observing platform. The entire assembly spun slowly around its central axis to create artificial gravity in the rim, which was the key innovation of an already innovative design. There were even elevators to carry crewmen from hub to rim and back. This was an astounding set of ideas, conceived at a time when most airplanes were still covered in canvas cloth and many people still drove Ford Model Ts.

The notion of a space station languished for a decade or two, but we know by now that people like von Braun and Oberth were thinking about space-related concepts all the time. In 1951, von Braun would inspire America with visions of an Earth-orbiting station in a major fashion.

Fig. 6.2. Wernher von Braun poses holding a model of his 1954 design for a space shuttle, with his "wheel" space station design in the background. Image from NASA.

In that year, a series of articles about space travel appeared in *Collier's* magazine, which in its day was like a literary Pinterest. Von Braun worked with his old friend Willy Ley,[6] another German space enthusiast, who, while scientifically trained, had turned his energies toward writing. With Ley taking authoring credit, an installment titled "A Station in Space" described von Braun's ideas about a space station, supplemented by lavish illustrations by famed space artist Chesley Bonestell. The series was a sensation and provided an optimistic and convincing story of a space exploration future that seemed inevitable.

Central to the article was a wheel-shaped space station designed by von Braun but identical in many critical details to Potocnik's, though more than twice as big. Von Braun's station would be 250 feet in diameter, house eighty crewmen, use solar power collectors along the rim, rotate at 3 rpm to produce artificial gravity . . . and be made of rubber.[7]

That's right. America's greatest rocket scientist proposed a 250-foot, triple-decked, solar-powered space station, and it would be the equivalent of a giant space-borne inner tube, made of soft, inflatable material, specifically reinforced nylon fabric. This material made sense, because even with von Braun's ever-larger rockets, and certainly with anything foreseeable in 1951–52, the station would have to be launched in small, light sections and assembled in orbit. Launching metallic segments of a 250-foot wheel in huge, macaroni-shaped sections would have been daunting and expensive, but sending up much smaller and lighter payloads consisting of a folded cloth section of the wheel, which would be inflated and then connected into a complete circle, made more sense.[8]

While von Braun saw a space station as just one component of his larger space exploration infrastructure that would culminate in *The Mars Project* scenario, he was careful to include a strong science instrumentation package. He was an engineer, yes, but also a scientist, an explorer . . . and a savvy politico.

The *Collier's* articles discussed onboard science labs that would contain "powerful telescopes attached to large optical screens, radar-scopes and cameras to keep under constant inspection every ocean, continent, country, and city."[9] Not only would this provide good science and reap benefits fitting the $4 billion investment (von Braun the politician again), but would, via continuous observation in a 1,070-mile-high polar orbit, make it "impossible for any nation to hide warlike preparations for any length of time. . . ." (von Braun the military pragmatist). The station had something for everyone—especially an American defense establishment nervous about the activities of the Soviet Union.

Von Braun's expansive prose went further: "Within the next 10 to 15 years, the earth will have a new companion in the skies." He continued, speaking of its utility: "A trip to the moon itself will be just a step, as scientists reckon distance in space." Of additional science: "There will be a space observatory, a small structure some distance away from the main satellite, housing telescopic cameras."[10] This was to prevent the motion of crewmen walking about his inflatable space station disturbing the sensitive telescope.

In a final note about military utility, von Braun added: "There will also be another possible use for the space station—and a most terrifying one. . . ."[11] And he went on to describe its utility for dropping nuclear bombs on enemy cities from space.

He concluded, with some certainty, that "Development of the space station is as inevitable as the rising of the sun." He was not wrong, but it would be decades before that dream would be accomplished, and in forms vastly unlike the 1952 design.[12]

As the 1950s drew on, the Potocnik/von Braun wheel-shaped station permeated popular thought about space stations. The *Collier's* articles, countless books for adults and children, and movies like *The Conquest of Space* by special effects guru George Pal meant that the design was stamped into the minds of the public. Oh, and there was one more person who might have had something to do with that . . . the inimitable Walt Disney.

A couple of years after the *Collier's* articles, von Braun was approached by the Disney organization to work with them on a series of programs for their new TV show *Disneyland*, later known as *The Wonderful World of Disney*. Von Braun and Disney's director of animation, Ward Kimball, produced three episodes on humanity's future in space, with the final episode airing in 1957—the same year that Sputnik launched. This Disney period was perhaps the ultimate rehabilitation of von Braun's image in the US. There had been much concern over his WWII affiliations with the V-2—visions of burning cities and soot-faced, crying orphans (enhanced by the powerful newsreels of the 1940s) were not far in the past, and there were many who disliked the idea of German refugees running America's rocket program. Von Braun was emblematic of the bunch to most Americans. But his appearances on the Disney TV show, complete with large and beautiful scale models of rockets and the space station, and his charismatic cohosting of the episodes, went a long way toward cleansing public perception of the former SS officer. It also further welded the idea of a future "wheel in space" into the minds of millions.[13] These productions were of the highest caliber and are still compelling to watch.

Other inflatable, wheel-shaped designs were studied well into the 1960s. A few full-sized mock-ups were even made by the contractors who, it was expected, would eventually build such a station. The designs varied widely; some were fatter and smaller, some folded up as a single mass with vertically hinging segments of the wheel's "rim." The notion of a centrifugal wheel was still in vogue when Stanley Kubrick's *2001: A Space Odyssey* premiered in 1968, with its enormous, rigid, twin-wheel space station dominating the poster art and early scenes of the film. This was the most spectacular version of the "wheel in space" vision, and the last. By this time, the Apollo program follow-ons included a well-planned version of Skylab, which would be constructed from a repurposed Saturn V third stage fitted out as a space station and launched in 1973. It turned out to be a spectacular success, but was a far cry from the large, expensive military stations planned in the 1950s.

Besides cost considerations, there were concerns about the use of inflatable structures in space, and despite much ground testing of scaled-down and full-sized versions, it was impossible to say how they would behave once in orbit until the first unmanned inflatables were sent aloft in the 1960s. Micrometeor impacts were one perceived threat; even a small impact could seriously puncture the structure. But as it turns out, there is far more to fear from the millions of bits of orbital junk launched by the space programs of Earth than from stray bits of rock and sand already in space. Overall, impacts and punctures have turned out to be a smaller risk than thought.

There was also worry expressed about the integrity of the structure from inside. One contractor study discussed the possibility of an overly active astronaut, sans space suit, losing his footing and plummeting through the reinforced nylon wall, dooming him to instant death in the vacuum of space and leaving a gaping hole in the station, which would depressurize quickly. While this scenario was extremely unlikely, the contractor was covering all the bases (the hyperactive spaceman would most likely have simply bounced off the hull, perhaps starting a spirited game of "bounce the spaceman off the wall").

NASA continued to look at inflatable structures throughout the decades, but never deployed anything man-rated in space. The copious research notes were publicly available, however, and a highly driven man named Robert Bigelow became interested and founded a company to explore the potential of inflatable habitats in 1999. He has to date invested somewhere between $250–$300 million of his own fortune, made from his ownership of Budget Inns of America.[14] Unsurprisingly, his primary motivation was to create orbiting hotels, but he has also closed the circle by working closely with NASA. In 2006 and 2007 he paid the Russian space agency to fly two small prototypes into orbit, called Genesis 1 and Genesis 2, and they performed well over two and a half years of testing. In 2016, a Bigelow habitat dubbed BEAM (Bigelow Expandable Activity Module) was flown to the International Space Station by a SpaceX rocket, attached to a station node, and inflated (NASA says "expanded") for research and evaluation purposes.[15] It's a smaller unit than his flagship BA330 modules at just thirteen feet long and ten feet wide (the BA330 will be fifty by twenty-two feet), but will provide proof of concept via careful monitoring by the astronauts aboard the ISS. They will not live or work in the module for now, but will simply enter it every few months to evaluate its integrity over time. The Bigelow expandable modules are a clever expression for the idea of inflatable habitats in orbit, and are being evaluated for use on the moon, Mars, and in interplanetary transit. It took over sixty years for von Braun's idea of an inflatable, manned structure in space to become a test-item reality, but it appears that the idea is here to stay, and may prove an invaluable addition to opening the space frontier.

CHAPTER 7

VENUSIAN EMPIRE: NASA'S MARS/VENUS FLYBY ADVENTURE

UNCLASSIFIED

In the mid-1950s, with the ravages of WWII fading to dull embers in the memories of most of the West, the future looked bright. Europe and Japan were rebuilding their economies, American industry and research were at an all-time high, and its postwar middle class was ascendant. First-world countries saw more consumer goods available to people of moderate means than at any time in history, and anything seemed possible. Even sending people to Venus.

Today, most people who paid even moderate attention in middle school science class know that Venus is not a prime vacation destination. It's a desolate, tortured, and blazing hellhole of a place with corrosive air at crushing pressures. The surface is an endless expanse of shattered gray rock, the skies a blank slate of permanent cloud cover, with a year-round temperature of about 900°F. Not only would your mai tai evaporate instantly, but if you had reason to bring along a block of lead, you could watch it slowly melt into the parched surface. Not that you'd live long enough to.

But in the anything goes, no-mountain-too-high, no-planet-too-far, strap-on-the-rocket-and-blast-off mindset of the 1950s, far less was known of the planet, and expeditions to fly past or orbit Venus seemed like a possibility. Little was known of the surface conditions there—the perpetual, dense-clouded skies did not permit visual astronomers to chart anything other than the cloud tops. So while giant telescopes revealed much about the surface of Mars (though the astronomers' interpretations were inaccurate—recall the notions about Mars during the planning of von Braun's *Das Marsprojekt* from chapter 3), our understanding of conditions on Venus was largely the result of indirect observation and guesswork. Most people, including many scientists, speculated for decades that nearby planets, the so-called terrestrial worlds (with the exception of sun-blasted Mercury), had atmospheres vaguely reminiscent of Earth. Allowing for temperature differences, why shouldn't similar conditions persist on Mars and Venus? Since Venus was cloud enshrouded and nearer to the sun, it was likely a hot, wet version of our own world—possibly reminiscent of Earth in its warmer, wetter past. It was not until 1958 that radio astronomers got a good indication of the surface temperature of Venus, and it was a shocker—their readings indicated about 600°F. Later measurements from Soviet landers pegged it at over 900°F.

But in the mid-1950s this was still not known, so it made sense to come up with some kind of plan to explore Venus the way we were considering the exploration of Mars. Perhaps not a landing, mind you, at least not right away, but a manned trip around Venus was something that might be accomplished within a reasonable time given the rockets we had on the drawing boards. Scientists and engineers in both America and the USSR had similar thoughts . . . as did at least one Italian.

In 1956, at an international gathering of spaceflight thinkers in Rome, an Italian rocketry pioneer, Gaetano Crocco, addressed the assembled luminaries with just such an idea. Crocco was born in 1877, about the same time that Giovanni Schiaparelli was making observations that would lead to his influential maps of Mars. By the 1920s, Crocco was following a course not dissimilar from the one of Robert Goddard in the US and, later, von Braun in Germany, building small liquid-engined rockets and firing them to ever greater altitudes in Italy. In the 1950s, possibly inspired by writers such as Arthur C. Clarke, who had written in broad terms about how a spacecraft might fly to Venus, Crocco worked out a possible mission design to send humans to the nearby planet. At the conference, Crocco outlined a mission design that would dispatch a manned spacecraft on a long, looping trajectory to fly past first Mars, then Venus, before returning to Earth. The spacecraft's departure from Earth would, he calculated, provide enough velocity to reach Mars, swing past the planet, drift back past Earth orbit, then sling past Venus and use that planet's gravitational influence to bend its trajectory back toward a rendezvous with Earth. The journey would take about a year and allow humans to visit two planets in one trip. The soonest possible alignment of the bodies to support such an undertaking would occur in 1971, according to Crocco.[1] This was an attractive alternative to another Mars-focused mission design that would have had a spacecraft head off to Mars, orbit (but not land) for 425 days, then head back to Earth. Besides adding a second planet to the itinerary, it would require roughly half as much fuel to accomplish, and get the crew back much sooner. Case made.

This paper would be pulled out and reexamined at periodic intervals, often alongside the earlier work of Walter Hohmann, a German engineer (and contemporary of Crocco, born in 1880) who published the first definitive study of low-energy routes between the planets in 1952, which have ever since been referred to as Hohmann transfer orbits. In both cases, an object leaving Earth would apply power to place it into a long, arcing trajectory toward another planet. Crocco elegantly added the components of gravitational effects of Mars and Venus to sling the spacecraft from planet to planet, ending up back at Earth.

These ideas of spaceflight, in particular Crocco's—which was labeled the "Grand Tour" (a term later applied to a couple of stillborn NASA mission proposals)—were reexamined in the 1960s. In 1962, the Future Projects Office (FPO) at NASA's still-new Marshall Space Flight Center (MSFC) granted funding to three major US aerospace contractors to study such a mission. Each was given a different mandate—General Dynamics (the parent company of General Atomics, recently of the Project Orion studies) was tasked with designing a Mars orbital or flyby mission, Lockheed would explore the same, and the Aeronutronic Division of the Ford Motor Company would study the Mars/Venus flyby mission design.

Referring to these types of mission proposals, one engineer at NASA's MSFC, Harry Ruppe, put it like this: "From the lunar landing in this decade to a possible planetary landing in the early or middle 1980s is 10 to 15 years. Without a major new undertaking, public support will decline. But by planning a manned planetary mission in this period . . . the United States will stay in the game."[2] Though he miscalculated the degree of political support in the US for reaching beyond Apollo toward the planets—essentially nil—he was right about public support for space exploration declining as the lunar program wound down.

NASA's name for the combined effort was EMPIRE, an overbearing-sounding acronym that stood for Early Manned Planetary-Interplanetary Roundtrip Expeditions. This plan for interplanetary flight was ambitious, but NASA had reason to be optimistic—the government's allocations for the manned spaceflight program were increasing every year.[3] President Kennedy had announced the moon as a goal just a few months earlier, and surely no country that would make such a huge, deep-pocket investment in space exploration would simply toss aside all that hard-earned technology, know-how, and infrastructure after accomplishing what many in the field viewed as an intermediate goal of reaching the moon. From that viewpoint, it does seem incredible that the Apollo flight hardware—the massive Saturn V, the Apollo command/service module, and the lunar module—was abandoned shortly after the lunar goal was met (it did contribute to two additional programs, the Skylab space station in 1973 and the Apollo-Soyuz orbital linkup in 1975). And there is, of course, the irony that the United States currently pays the Russian space agency to ferry American astronauts to the International Space Station aboard the Soyuz, hardware that was designed for their competing—and ultimately unsuccessful—lunar effort.

All of the EMPIRE proposals were to be based, whenever possible, on technologies then being researched by NASA and, specifically, von Braun's group. These proposed technologies included rockets such as the Saturn V and some larger than the Saturn V—one was called Nova, a beast of a rocket that would use eight F-1 engines in the first stage (the Saturn V would utilize five F-1s), and the Supernova, an even larger rocket. But the EMPIRE studies pushed the envelope, and included nuclear propulsion units that were also being studied (and would later be successfully tested). These units would provide long-duration rocket engines that would use a fission reactor and hydrogen fuel to propel the spacecraft after it reached orbit, using conventional rockets (not an atomic sendoff like the Orion program had envisioned).

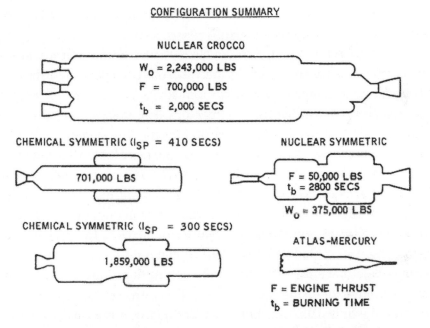

CONFIGURATION SUMMARY

NUCLEAR CROCCO

W_o = 2,243,000 LBS
F = 700,000 LBS
t_b = 2,000 SECS

CHEMICAL SYMMETRIC (I_{SP} = 410 SECS)

701,000 LBS

NUCLEAR SYMMETRIC

F = 50,000 LBS
t_b = 2800 SECS

W_o = 375,000 LBS

CHEMICAL SYMMETRIC (I_{SP} = 300 SECS)

1,859,000 LBS

ATLAS-MERCURY

F = ENGINE THRUST
t_b = BURNING TIME

Fig. 7.1. Schematic comparison of Ford Aeronutronic's designs for a Mars/Venus flyby spacecraft. The center right version, labeled "Nuclear Symmetric," was by far the lightest and most efficient version. Image from NASA.

In the introduction to one of the two plans submitted by Aero-nutronic, "The EMPIRE Dual Planet Flyby Mission," the authors said of NASA, "Much credit must be given to the forward thinking approach given to the NASA on this program in 1962. By attacking the areas of interest at this early date, it was possible to obtain a clearer picture of the requirements for early manned planetary and interplanetary flight. Thus, the nation's resources, and the NASA and other United States space programs, can be oriented toward long range goals at an early date."[4] If only he'd known that the Apollo landings would be the last gasp of human exploration beyond Earth orbit for the twentieth century (and, so far, beyond).

This version of Aeronutronic's plan, completed in 1962, presented a detailed accounting of an adjusted version of Crocco's slim document. It was assumed that by the time the spacecraft components would need to be launched for the 1970 departure window, NASA's heavy booster inventory would include not just the Saturn V, but also one or more variants of the larger Nova.

The preference was for the nuclear rocket propulsion system to power the in-space portion of the journey. NASA had been laboring over designs to put atomic power to work in space, not just for propulsion but for power-providing nuclear reactors. The nuclear rocket engine program was called NERVA (Nuclear Engine for Rocket Vehicle Application), and had begun early in the 1950s at the Los Alamos National Laboratory. NERVA was at the core of many of NASA's interplanetary plans for the future due to its ability to propel spacecraft much more quickly and efficiently across vast distances and over long periods of time via continuous thrust. Numerous tests were conducted in Nevada between 1959 and 1972.[5] One of these ran the engine for 115 minutes with twenty-eight startups—early proof of the firing durations and start/stop ability needed for deep-space missions. This promising program would be shut down in 1972 for budgetary and political reasons (launching nuclear materials into space—no matter how safely packaged—made people nervous). This was a great loss to American spaceflight efforts. Nuclear rockets had more than double the efficiency of chemical propulsion systems at roughly half the weight,[6] and could have done much to open the solar system to human spaceflight.

Aeronutronic studied a number of trajectory options, including revisions of Crocco's. In refining some of his calculations, the engineers slid the launch window and discovered that the journey would take a bit longer than Crocco thought, but the variations were minor— Crocco's trajectory looked to Aeronutronic like it would take 396 days. There was another trajectory they evaluated, however, called a symmetric trajectory. This flight plan would take longer—611 days, but would require less fuel, and the weight savings could more than make up for the extra mass required to feed and support the crew during the longer voyage. Whether or not the added time would have been a problem to humans in space is open to question—clearly the long-term exposure to both radiation and weightlessness would have been profound issues. But little was known of either of these risks at the time, and with this trajectory, a far smaller spacecraft would suffice.

To put it in terms that senators and congressmen could understand was easy: weight equals cost, and less massive spacecraft are much less expensive to launch than heavy ones. And if a mission can be mounted with a single launch, that's far better—safer as well as more affordable—than something that takes a series of launches and risky, in-orbit assembly of multiple components. The version suggested for the revised Crocco trajectory would be 1,121.5 tons of mass (with multiple launches and assembly in space); Aeronutronic's symmetric trajectory plan would require just 188 tons (a single Nova C-8 launch),[7] if nuclear engines were used for in-space propulsion. The new symmetric option looked better all the time—it was cheaper, would probably work, and was less risky—in all, sweet music to a politician's ears.

For either mission profile, Aeronutronic noted that life support and radiation protection would be of paramount importance. A longer mission would require more life support supplies, but only approximately the same radiation shielding. Regardless of mission duration, radiation "shelters" for riding out high-energy events such as solar storms would have to be provided, and the overall design of the crew quarters would need to offer some level of shielding as well. Even so, the longer-duration symmetric plan still won out in terms of overall mass and cost efficiencies.

The proposed spacecraft was a 156-foot long, 33-foot-diameter beast. Its single nuclear engine would be capable of about 50,000 pounds of thrust—about two-thirds that of the Redstone rockets that sent the first American astronauts into space. But its capability to fire for very long periods of time more than made up for the modest output. Departure from Earth orbit would need about fifty minutes of firing time, or well less than half the duration of NERVA's ultimate firing tests. The output figure for the nuclear engine was later revised to an "ideal" thrust of 200,000 pounds, equivalent to the chemical rocket engine used on the Saturn V's S-IVB stage (which propelled the Apollo missions out of Earth orbit and toward the moon).[8]

Much of the spacecraft's mass would be the large tanks of liquid hydrogen fuel for the NERVA engine. When the nuclear rocket completed its initial firing to depart Earth orbit, the first batch of empty fuel tanks would be released and sent adrift. Between fuel and tank mass, this would shed almost sixty tons. A second firing would deplete another thirty-eight tons of fuel, but when it was over the empty fuel tanks would be retained as part of the radiation shielding (an additional twenty inches of plastic sheathing would provide protection from solar radiation). Finally, with all fuel depleted, the nuclear engine itself would also be cut adrift. The craft would now be down to about seventy-eight feet in length.

Included in the design of the long-duration spacecraft were a command center, an experimental lab (research into various zero-g topics would be conducted en route, just as they are aboard the ISS today), and the radiation shelter. Two additional small crew modules would be extended on retractable arms, and the ship itself spun along its long axis to create a low-RPM centrifuge to assert partial gravity in these areas for the duration of the voyage, thought to be critical to the crew's health.[9] At the back, near the NERVA engine mount, would be one or more small nuclear reactors for supplying power to the spacecraft.

If the trip had begun at the right time in mid-1970, the Venus flyby would occur first, after about one hundred days in space, at an altitude of about 4,900 miles (Mariner 2 had passed the planet at a distance of 21,600 miles in 1962). Then, 195 or so days later (traverse times depended on the launch date), the spacecraft would sling around Mars. The altitude of this flyby could be as low as 2,740 miles (Mariner 4's closest approach in 1965 was about 6,100 miles), which would allow for some detailed observations; though most of what they would have been able to accomplish has long since been surpassed by

NASA's unmanned Mars probes. Some 312–340 days later, the spacecraft would return to Earth.[10]

The crew would transfer to a return vehicle to return to Earth's surface. While NASA had stated a preference to utilize as much Apollo hardware as possible, the first iterations of Aeronutronic's design called for a lifting body to return the crew, not unlike the air force's LUNEX Earth Return Vehicle. It was thought that the "winged" vehicle would give the crew more options for a safe reentry than a capsule. Once aboard the ERV and separated from the now-unneeded interplanetary craft, the crew would fire braking rockets and descend back to Earth.

The Mars/Venus flyby mission as laid out by Aeronutronic was estimated at about $12.5 billion.[11] This does not seem like an outrageous price, given that Apollo cost between $20–$24 billion.[12] But it is likely that any mission like EMPIRE would have engendered cost overruns, possibly vast ones, and schedule delays. Of the latter, it was suggested that nuclear rocket engine development be started within six months to keep things on schedule—sooner if possible: "It would appear that even now, a launch in July 1970 is within technological reach of a plausible development program—provided certain technological problem areas are attacked immediately."[13] NASA did mount the "attack" suggested, but just not to the planned interplanetary voyage. Their efforts went into the increasingly demanding Apollo lunar landing program instead. The only major follow-ons to Apollo would be flown in Earth orbit: Skylab and Apollo-Soyuz.

Of course, over all of these grand interplanetary plans hovers a large question: why send *people*? We've seen one incredible success after another in robotic exploration since 1962, and while (from today's standpoint) it's vastly sexier to see booted and backpacked astronauts planting a flag on another world, is it really necessary? Especially considering that for this part of the EMPIRE plan, we're talking about flybys and not landings?

As seen through the eyes of a spaceflight planner in 1961, it made sense. Robotics was in its infancy, and robotic machines capable of flying, even on a larger rocket like the Saturn V, were not a proven quantity.

Computers were huge, transistors were just coming onto the scene, and controlling robots at the vast distances required was a virtual unknown. While there were some early and notable successes—the first Sputniks and US satellites—their robotic functions were minimal. The first human spaceflights used mostly electromechanical controls, often operating with ground-based prompts, with only rudimentary automation. The first successful US planetary encounters—Mariner 2 at Venus and Mariner 4 at Mars, which required some level of robotic self-sufficiency—were just getting underway. All this is to say that it was unknown how well these machines would perform at planetary distances. A better way to proceed, they thought, would be to send human crews past Mars and Venus on flyby missions, and they could dispatch robotic probes while they were in the neighborhood (as they looped the planets at close range). Humans controlling those machines while in close proximity and with little to no radio delay should be far more reliable.

Of course, as we have seen, the robots do just fine. In today's world, where the Curiosity Mars rover can land itself and drive for a day or more unassisted, and Elon Musk's rockets can make autonomous, pinpoint landings on a lurching seagoing barge after reentering the atmosphere, the mission plans of EMPIRE seem quaint—a product of a bygone era, notions of romantic human exploration in deep space.

Even at the time there were detractors. Max Faget, the designer of the Mercury, Gemini, and Apollo capsules as well as a major contributor to the space shuttle, demurred. He felt that the "overall planning of a total spaceflight program should be based on a logical series of steps." Faget felt that the flights of Mercury and Gemini were engineering development missions, after which the Apollo spacecraft would engage in "the first real mission." After these increments might come a space station and lunar base, the inference being that planetary flybys were a stunt.[14]

As Aeronutronic was delivering its audacious plans for a two-year journey to Venus and Mars, the other two contractors completed their work as well. For well under a million dollars in seed money, NASA now had three major proposals, with multiple options for each, on the table. Once the moon had been reached and possibly explored with an extended stay program, nearby planets would be ripe for human exploration. Of course, as we know, no such thing occurred.

EMPIRE, in any of the versions delivered by Aeronutronic, General Dynamics, or Lockheed, was doomed before the reports were sent to NASA. Not that it wouldn't have worked. Some of the major components—notably the Saturn V and the NERVA engine—were well into development. No, what was missing was willpower in Congress and the executive branch to go beyond Apollo and to create a long-term space exploration infrastructure with the incredible advances of the early 1960s. It seems fair to say that the government—and perhaps the bulk of the American populace—experienced a "failure of imagination," as one of the Apollo astronauts would later opine about early shortcomings in the lunar landing program.[15]

Despite this, the idea of planetary exploration, even as a flyby, did not die so easily. As we will see in later chapters, the idea of a crewed expedition to land on and explore Mars has continued to this day. As Mars's secrets became better understood with robotic probes delivering streams of data, plans for crewed interplanetary spaceflight have been redesigned and updated by NASA every few years.

Human flights to Venus grew less appealing, however, as the truth about the planet became known, and robotic spacecraft were proved adequate to the task. While the Soviet Union had virtually no luck at exploring Mars (and not for lack of trying—they have sent scores of probes to the red planet and to date none have completed their mission), Venus was another story. Russia's attempts to reconnoiter the planet began about the same time as NASA's, in 1961–62. The first US mission, Mariner 2, succeeded but returned limited data due to its being a rudimentary spacecraft.[16] But in 1967, and again in 1969, the Soviets were very successful at Venus, first at reaching and entering the atmosphere, and, in 1969, landing an operational probe on the surface. The lander did not last long, but transmitted the harsh truth to the world: Venus was a hellish place.

Regardless of these events, there was still some interest in a manned Venus flyby. Keeping the idea alive was a challenge, since the robots were now smart enough and (mostly) reliable enough to explore the other planets without on-site human control, and by 1966 NASA's orders with prime contractors for Apollo hardware were set. The command modules and lunar modules needed for the lunar land-ings would require work right up until the last moment, and construc-tion of the Saturn V was essentially complete. No new large-scale spacecraft orders were coming from the Johnson administration.

The last straw the contractors and NASA could clutch at was a Mars and/or Venus flyby using Apollo hardware. There would be no Nova booster (it had not been needed since the decision to use the lunar orbital rendezvous approach for Apollo), no gliding Earth Return Vehicles, and no nuclear rocket engines. We had Saturn Vs, command/service modules, as well as a flexible architecture in the Saturn's third stage, the S-IVB. This stage, the top stage on the Saturn V's lunar configuration or "stack," would ultimately be used to build the Skylab space station in 1973. Why not use it to travel farther into interplanetary space? Or, for that matter, custom hardware could be made to fly atop the Saturn. The architecture of the Apollo hardware was robust enough to accommodate a number of design choices. Perhaps there would be some hardware leftover after the lunar missions were accounted for, or they might be able to squeeze enough money out of Congress for a few more rockets....

A spate of planning for Mars/Venus flybys, orbital missions, and (in the case of Mars) some landings (both manned and robotic) fol-lowed. If you could stack the paperwork, you might be able to climb to the red planet. From today's perspective, it appears as an almost equal balance of optimism that our future in space was bright and vibrant—just look at that amazing Apollo hardware!—and a desperate attempt to keep that same Apollo machine running, but the layoffs from Apol-lo's assembly lines were not far off. In either case, the Apollo Applica-tions Office at the Johnson Space Center, the Future Projects Office at the Marshall Space Flight Center, and many contractors were engaged in the invention of a vibrant future for Apollo that, beyond Skylab, would exist only on paper. Most efforts for interplanetary journeys were focused on Mars flybys and landings, but a few were specifically targeted on the red planet's less glamorous cousin, Venus.

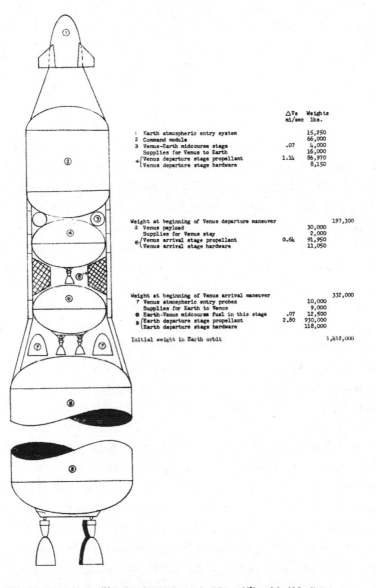

		ΔVs mi/sec	Weights lbs.
1	Earth atmospheric entry system		15,250
2	Command module		66,000
3	Venus-Earth midcourse stage	.07	4,000
	Supplies for Venus to Earth		16,000
4	Venus departure stage propellant	1.14	86,970
	Venus departure stage hardware		8,150

Weight at beginning of Venus departure maneuver — 197,300

5	Venus payload		30,000
	Supplies for Venus stay		2,000
6	Venus arrival stage propellant	0.64	91,950
	Venus arrival stage hardware		11,050

Weight at beginning of Venus arrival maneuver — 332,000

7	Venus atmospheric entry probes		10,000
	Supplies for Earth to Venus		9,000
8	Earth-Venus midcourse fuel in this stage	.07	12,500
9	Earth departure stage propellant	2.80	930,000
	Earth departure stage hardware		118,000

Initial weight in Earth orbit — 1,412,000

Figure 6 - Typical Space Vehicle; 565 Day Venus Orbiting Stopover Round Trip in 1980, e= 0.9. 40 Day Stay at Venus, Type II-D Trajectory. Apollo Level of Propulsion Technology. See Tables I and II for Other Inputs.

Fig. 7.2. Schematic view of a later NASA study for a Venus orbiter, using mostly Apollo flight hardware as a basis. At top is the crew reentry spacecraft, and below that (labeled "2") is the crew module, home for the duration of the mission. The rest is comprised of propellant tanks. Image from NASA.

A follow-on study was conducted by MSFC in 1965. This plan specifically required the use of Apollo-based hardware to make the Venusian trek, a smart move by MSFC since much of the Apollo hardware was being built under its purview—missions such as this would help keep the assembly lines open and the furnaces stoked. Like EMPIRE, both Mars and Venus missions were considered.[17] In the case of Mars, the flyby was to be a manned "reconnaissance" of the planet that could help in mission planning for a landing in the 1980s. Venus was a target of interest on its own merits—that is to say, no landings were envisioned; it would just be a fascinating place to fly past and investigate more closely.

Among the options considered in this 1965 report were missions to Venus intended to take place in the late 1970s after the end of the Apollo landings (which so far as anyone knew at the time could go up through Apollo 20).[18] A few options were considered, from 661 to 691 days, including both Mars and Venus flybys. The study added that since a manned Mars landing mission would probably require something beyond chemical propulsion, such as nuclear rockets (a subject taboo to mention in a near-term context), flybys of Mars and Venus, using Apollo-derived technology, would be a way to keep the public engaged in the interim.

In mid-1965, in this atmosphere of some optimism—and with Mariner 4 set to return science results from Mars soon—the National Academy of Sciences convened a group called the Space Science Board in Woods Hole, Massachusetts. Part of their working mandate was "first, to develop a program of planetary exploration and to recommend priority within it; second, to determine the need of astronomy in space; and, third, to consider the role of man in space research."[19] The group's conclusion was that NASA should concentrate its efforts after Apollo on planetary exploration—in effect, a mandate to explore the two nearest targets, Mars, the moon, and Venus, in that order. With plans for the moon moving apace, the groundwork had been set out for continuing studies of how to reach the other two worlds. This matched MSFC's agenda nicely and set the stage for more studies.

One of these subsequent studies—and one of the more compelling plans—was delivered in a 1967 study by NASA contractor Bellcom, working with MSFC and the Johnson Spaceflight Center (JSC) in Houston. The Bellcom study suggested a voyage to begin in 1977 that would send astronauts in almost entirely Apollo-sourced hardware past Venus, then a loop past Mars, and back past Venus before coming home.

The spacecraft would have a mass between 145,000–150,000 pounds—not much larger than 1973's Skylab space station—and carry a crew of four to six men. Life support for two years' duration was specified.

After their 1977 launch atop a single Saturn V rocket, they would complete the Venus/Mars/Venus/Earth trajectory in just under twenty-four months. Their first pass of Venus would occur 149 days after launch, and a suite of probes and instrumentation would be released. These probes would include a series of balloons that would inflate while in the upper atmosphere to send back atmospheric data. A Venus orbiter would be dispatched, and "sinkers" (slow-parachuting cameras) and "impactors" (hard landers), designed to take images of and investigate the surface, would also be deployed—it was hoped that an impactor might be able to survive for at least an hour to transmit data from the surface. The goals of the Venus investigations were summarized thus: "Venus . . . is a cloud-covered planet with a dense atmosphere and a hot surface. Also, it is roughly the size of the Earth, with a Venus day being equal to about 120 Earth days. This sort of picture is not readily conducive to thoughts of a manned Venus landing. Instead, the reaction is one of probing the environment more thoroughly to understand the basic physics and chemistry (and possibly biology) of the planet."[20]

Upon reaching Mars in mid-voyage, some 349 days into the voyage (almost 200 days after the first pass of Venus), three landers would be released to head toward the surface. The idea of a sample return was mentioned, specifically to search for life there (this is something we are still trying to achieve robotically today). The landers would reach Mars ahead of the crewed flyby module, collect their samples, and send them back into space to rendezvous with the passing manned craft, which would carry them back to Earth (the samples would also be examined, in at least a rudimentary fashion, in an onboard lab).

The spacecraft would then swing by Venus a second time 574 days after leaving Earth. A second Venus orbiting relay satellite would be held in reserve, and the crew's activities there would be dictated to a degree by the results from the first batch of probes released earlier. It was assumed that the impactors would be long dead and the balloon instruments past their operational lifetimes, and the second orbiter would be utilized to relay data from new balloon probes launched on this pass of the planet.

A total of four Venusian balloon probe assemblies would be launched, with six balloons deployed from each. Two orbiters would relay data back to the spacecraft. Eight Venus landers and eight "sinker" probes—machines that would transmit images and send data as they slowly descended—would also be deployed. The Martian leg of the journey would require three lander/sample return machines.

A number of asteroids would be visible from the spacecraft's trajectory for telescopic study—and Mercury, along with Mars's moons, would be investigated telescopically, as well as the sun and various deep-space objects. These activities would certainly help offset the ennui of such a lengthy journey.

Finally, in early 1979, after 716 days in space, the astronauts would return to Earth in an Apollo command module, toting samples of the Martian surface and reams of data from both planets, multiple asteroids, and other space-voyage phenomena encountered along the way.

Like other similar mission designs, the point was made that once the spacecraft was boosted away from Earth, only midcourse corrections would need to be made; otherwise, it was a free ride all the way home. The planetary flybys would provide the needed course changes to return to Earth after the long journey.

The many proposals received similar treatment when they were submitted and reviewed—thanks, but no thanks. Even before most of these reports were received, and certainly well before the projected flight dates, the Johnson administration had signaled its lack of support for extended deep-space exploration. An expensive lunar landing program, the seemingly endless Vietnam war, and other political considerations were at the root of this decision.

It should be noted that certain assumptions had to be made to budget such mission designs. In the case of studies that came out of NASA, particularly from MSFC, von Braun and his team had great plans for the Saturn V. In some of the studies performed in the mid-1960s, his "per-flight" cost of a Saturn V was based, by necessity, on how many of the giant rockets were built and flown—the more they made, the cheaper they became. Dozens of the giant rockets were foreseen.

As it turned out, by 1973 (when the last Saturn V flew for Skylab), only fifteen had been built and thirteen launched (the remaining two are in museums). Depending on how you calculate the costs, the price per flight was somewhere between $185–$500 million. But for some of the deep-space mission studies, like those discussed in this chapter, von Braun assumed that by the mid-1970s, more than sixty Saturn Vs would have launched, having reduced per-flight costs to about $60 million (or about what Elon Musk charges in today's dollars for a Falcon 9 launch). Fifty-two or more smaller variants of the Saturn, the IB, would have launched, dropping its per-flight cost to $22 million. And what of the brilliant command/service module from the Apollo lunar flights? We would have flown seventy or more of those as well, dropping the unit cost again to the $70 million range per unit.[21] Human spaceflight was beginning to look like a pretty good deal when flown in quantity.

Of course, we never came close to that flight rate. But, looking again at planetary flights, there was at least one more arrow in Venus's quiver. How do you add value to a planetary flyby mission? For starters, how about finding a way to stay there for a while? For flyby trajectories, after all that time crossing the void of space, you have about ten hours of looping the planet at closest approach—hardly a robust payoff for such an arduous journey.

A 1967 paper from Ed Willis, an engineer at NASA's Lewis Research Center, defined a minimally demanding program for a Venus orbiter.[22] It too would be based on Apollo hardware, and by using a long, elliptical orbit he felt that his stay-at-Venus mission design would be much less demanding in terms of weight and cost than the proposed Mars/Venus flyby tours. Willis further expressed that Mars missions should be about surface landings and exploration, and Venus made sense for observation from space . . . and that only an orbiting mission would provide the scientists aboard enough time to do their work properly. The departure dates ranged from 1976–1986, and flight durations varied. An indicative example would be a 565-day mission, including forty days in orbit around Venus. But the range of his calculations gave many more choices:

> Venus is our nearest planetary neighbor and is an interesting object for scientific exploration following the manned lunar landings in the early 1970s. In this paper manned orbiting stopover round trip missions to Venus are studied for the 1975 to 1986 time period for trip times ranging from 360 to 660 days, with stay times up to 100 days, for Venus parking orbit eccentricities from 0 to 1.0, and several levels of propulsion technology.[23]

Willis did toss in an argument for a nuclear engine: "While the Venus orbiting mission can be accomplished using the Apollo level of technology, reductions in weight are possible using advanced propulsion. [This kind of] departure maneuver can reduce the initial gross weight by 30 percent. . . ."

And, in a final bid to give favor to Venus over Mars as a solo target: "To accomplish a Mars orbiting mission in the easiest year would require a vehicle 70% heavier than that for the Venus orbiting mission in the most difficult year. The disparity can be much larger in other years."

Alas, for either Mars or Venus, flybys or orbits, the door had slammed closed. Whether born of magic or geopolitics, the moment that had spawned the Apollo program had passed, and after 1972, humans would not leave Earth orbit again. But we certainly could have, and a Mars/Venus flyby mission would have been a wonderful adventure.

CHAPTER 8

BLUE GEMINI: WEAPONIZING ORBIT

CLASSIFIED: *DECLASSIFIED IN 2015*

D espite the stillborn Horizon, LUNEX, and other program proposals, departments of the US military continued to covet the
idea of a presence in space. Much of this hinged off German
plans from before and during WWII. From the robust V-2 program of
von Braun—to slam ballistic missiles into enemy cities—to more exotic
plans, such as Hermann Oberth's orbiting "sun gun" (as discussed in
chapter 6), the Germans had held an early lead regarding planned
space weapons. By the mid-1950s, numerous ideas were being formu-
lated in both the US and USSR for seizing the "high ground" of space,
be it Earth orbit or the moon.

The US Air Force had emerged post-WWII from the US Army Air
Corps. The army and navy had kept their aerial operations separate
throughout the war, though when the air force came into existence, the
navy kept its largely carrier-based aerial force for itself. This left the
three divisions of military authority within the US defense establishment—the army, to carry out land-based combat, the navy, to dominate
the seas and carry out ocean-oriented aerial operations (and the Marine
Corps within that organization), and the air force. Given these divisions
of responsibility, it seemed obvious to many in the new organization
they were the natural choice for space-based military operations. This
spawned the plans for LUNEX and other air force–only programs.

But when NASA was created as a civilian agency in 1958, it threw a
wrench into the air force's plans. President Eisenhower was not a fan

of militarizing space. He had been intimately exposed to the highest levels of managing the largest war in history, and saw no reason to extend manned military operations into orbit. The air force would simply now have to split the available money and technology with NASA if it was to have any role in space. This notion did not sit well with many of the generals who had experienced WWII and the Korean War. They wanted to control America's manned space endeavors directly.

Besides LUNEX, the air force had proposed a winged shuttle-type spacecraft for use in Earth orbit. This was a far smaller craft than the space shuttle that ultimately dominated US spaceflight from 1982–2011, and was designed with military objectives in mind. By the time NASA was officially formed, this spaceplane, dubbed Dyna-Soar and later the X-20, had been on the books for at least a year officially, and had existed far longer as an unofficial ambition (see chapter 12). It was a direct outgrowth of the X-15 program, and to men steeped in winged flight, a space-capable hypersonic glider made all the sense in the world. The ballistic capsules that NASA was suggesting for manned spaceflight seemed a distraction, a stunt almost, simply to outpace the Russians.

At the time, the challenges of operating winged craft at the edge of space were poorly understood, however, and the interim step— the X-15 rocket plane (which would reach the edge of space but not orbit) was just beginning. Ballistic capsules—simple, small one-man affairs—were the preferred choice of a large cadre of civilian engineers. And regardless of the air force's preferences, that program was moving forward aggressively. A capsule could also fly much sooner than Dyna-Soar, which would be flight-ready, at the earliest, in 1965. So, always able to switch to a strategic mindset, the blue suiters bit the metaphorical bullet. The immediate future would be skewed toward space capsules. In any event, Dyna-Soar would ultimately be canceled in 1963 after spending over $400 million in development,[1] with just the bare beginnings of construction underway.

After much internal debate, an if-you-can't-beat-'em-join-'em approach was adopted, and the air force began planning a series of orbital flights using NASA-designed hardware. By this time, the single-seat Project Mercury was nearing its end with the final flight occurring in 1963. Construction of NASA's two-seat Gemini program was underway, and this larger spacecraft had vastly improved capabilities. The driving force behind Gemini was research and experimentation with rendezvous and docking between multiple spacecraft—the Apollo lunar landing program would need this capability to complete its goals. If two craft were not able to find and connect with one another in space, the deal was off. Also, working in space—leaving the spacecraft and completing basic tasks—was a priority, as were extended operations in space, since the astronauts would have to be capable of surviving in that environment long enough to complete their lunar missions. In light of these requirements, the Gemini spacecraft was designed to carry two astronauts and had more advanced piloting and navigation capabilities. It also had a profoundly different set of trajectory-altering tools. While the Mercury spacecraft was capable of changing its orientation in space, Gemini was able to actually change its trajectory—in various guises, it could change orbits and head off to rendezvous with other spacecraft. It was closer to a two-seat fighter jet in capability than it was to Mercury, and that among other qualities made it appealing to the air force, which was at any rate running out of space-capable options. The air force would join with NASA (albeit with cautious reluctance on both sides of the equation) and stage a series of military flights with NASA hardware (which was built by McDonnell Aircraft, later to become McDonnell Douglas). The air force's program would be called Blue Gemini.

Behind the air force's decision was a member of the top brass of the air force's ballistic missile program, Bernard Schriever. A pivotal figure in the development of the Atlas ICBM, which also powered the orbital Mercury missions, Schriever was a natural choice to spearhead working with NASA. But his efforts in rocketry and missiles had long been dismissed by another air force general, Curtis LeMay, the heavyweight behind the United States' bombing programs of WWII and chief of staff of the US Air Force. It did not help that LeMay had once been Schriever's commanding officer, and, had he a choice, would not have given Schriever that fourth general's star. This acrimony would continue throughout the air force's involvement with spaceflight in the 1960s. By the time Schriever retired from active duty in 1966, he oversaw almost half of the organization's budget—surely a thorn in LeMay's side (LeMay retired in 1965).[2]

Schriever had an interesting history. He was born in Germany in 1910 (which may have accounted for some of LeMay's disdain),[3] and his family was able to emigrate to the US on the eve of America's entry into WWI (his father had been interned on a German ship in New York in 1916, and his mother followed later). Schriever served under LeMay as he piloted missions over Japanese-held targets in the Pacific theater in WWII. After the war he joined the new US Air Force, starting his work with missile and satellite programs in 1954.[4]

While part of the schism between LeMay and Schriever may have been over the fact that ballistic missiles were robotic devices—LeMay was an advocate of manned bombing, hence his strong preference for bombers—Schriever was more open minded. When ICBMs were first entering the scene, computers were large and cumbersome, and controlling a missile from launch to impact, thousands of miles away, was a daunting task for an onboard computer. It took years, but the flying calculators were eventually proved to be quite effective in targeted tests. However, Schriever was not anti-pilot—few former pilots would be. When the opportunity arose to place air force personnel into space-bound machines, he was among the first to embrace it. Blue Gemini seemed the perfect compromise.

It should be noted that while the air force was fond of winged spaceflight, this would not be its first foray into planning for flight in ballistic space capsules. Even as Dyna-Soar was ramping up in 1957, these men were considering a tiny effort they called "Man in Space Soonest" (MISS). Despite the unfortunate acronym, their goal was a reasonable one: do whatever it takes to beat the Russians into space. The MISS spacecraft would have been even smaller than the Mercury capsule (which was already incredibly cramped), basically a lone pilot wearing a metal cone. When space programs were being reviewed in the late 1950s, however, Eisenhower was critical of the overlap between Mercury and MISS, and the air force's program was canceled.

Fig. 8.1. The Gemini B spacecraft, part of the US Air Force's attempt at manned spaceflight. Despite years of planning and a logical flight manifest, an air force Gemini never flew. Image from USAF.

True to his pilot's instincts, however, Schriever would not let the work already done be put to pasture. He followed the cancellation of MISS by directing the contractor developing the capsule, the Lockheed Corporation, to convert the design into an unmanned satellite called Samos, with a large camera package installed in the pilot's seat. It flew, and failed, five times, but kept the air force in the spaceflight game. Interestingly, like many unmanned Soviet spacecraft, the instrument-carrying capsule was pressurized. The Soviets pressurized their robotic spacecraft because their electronics would only function reliably inside a pressurized vessel. The air force did not have the same requirements . . . so Schriever may have had other motivations.

NASA's Gemini would fly on another air force rocket, the Titan. This was more powerful than the Atlas used for the Mercury program and was needed to loft the heavier two-man spacecraft. In 1962 the Blue Gemini program was made official (though classified), and planned for up to seven flights. The air force also had ambitions of a space station, an orbiting military observation platform, but Blue Gemini would take place first—it was simpler and would get them into space two to three years sooner. They would launch the first military Gemini in 1965, the year following NASA's flight of the new spacecraft, and continue through the following year. Then the air force's orbiting lab—called the Manned Orbiting Laboratory—would start ramping up concurrently.

While NASA focused obsessively on reaching the moon, the air force was eyeing a long-term program of orbital observing—primarily spying on the Soviet Union. But interestingly, despite this primary motivation, the blue suiters' immediate flight goals overlapped NASA's nicely. Long-duration flight, orbital maneuvering, rendezvous, and docking were all shared desires of the program. A one-man jet pack, the Astronaut Mobility Unit or AMU, was also under development to allow individual astronauts to leave the Gemini capsule and fly freely, untethered, through space.[5]

The blending of air force and NASA objectives continued into flight planning. The first two flights of air force pilots would occur seated next to NASA astronauts, gaining experience in the spacecraft and supporting NASA's mission objectives. Then air force crews would fly the remainder of the "blue" missions, with both NASA and air force objectives included in the mission plan. And, doubtless to the satisfaction of the air force cadre, the Gemini capsules were originally designed to fly to a gliding landing, as opposed to splashing down in the ocean as Mercury had (and as all blunt-bodied US spacecraft have done).

This is an interesting sidetrack on its own, but in brief, these runway-landing Gemini craft would reenter like all capsules, back-end first, with the heatshield slowing the spacecraft as it descended from orbit. It would then be reoriented as the parachute was deployed— but this was no ordinary parachute. In a design innovation that must have lifted the air force's spirits, what popped out of the side of the hurtling Gemini capsule was an inflatable wing, a parasail. Called a Rogallo wing (after its inventor Francis Rogallo, who worked at NASA's Langley Research Labs), it was a delta-shaped, twin-humped fabric triangle with inflatable stiffeners on either side. With this para-wing, the capsule would fly horizontally, with the pilot looking out of the tiny, eye-shaped window in the Gemini hatch. Tricycle-style landing gear (like those used on a Cessna private plane, except with skids instead of wheels) would unfold from the bottom of the capsule, and it would be *flown* to a runway landing at Edwards Air Force Base in California (where the X-15 and the first shuttle flights landed). The word *flown* was an important distinction—Mercury capsules fell, under parachutes, affected by wherever the prevailing winds pushed them, to splash into the ocean where a number of navy ships awaited. Once they reached the water, the astronaut would bob around, waiting passively for a ship to come and recover him. To air force pilots the difference was huge—they would have control over their spacecraft at all times; no undignified floating like a cork in the ocean for them.

But after extensive testing and many tens of millions of dollars invested, the horizontal-landing Gemini system was just too balky and dangerous. The para-wing would often not deploy properly. Controlling it was difficult. The cone-shaped Gemini was not much of a glider, and handling the control system was tricky. And after two crash landings by test pilots, one of which involved a dangerous ejection and the other a very hard touchdown (which left the test pilot in the hospital for weeks), the horizontal landing plans were scrapped. It probably could have been made to work, but there was simply not time. The first Gemini flight was scheduled for 1965, and that date was immutable. The simpler parachute-into-the-sea system would have to do. Strike one for the air force's plans.

But the program continued, running quietly alongside NASA's much more public flight schedule. Other air force objectives included the building, flying, and testing of an inertial navigation unit, an inflight computer and tracking system that would track the location of a craft in space during flight without external input. Inertial navigation was important to both the air force and NASA, so this would be a shared win. Other experiments included a large ground-mapping radar (which would have a solo pilot, with the copilot's seat being filled with camera hardware) and an inflatable module that would test the use of nonrigid materials in space, primarily for an inflatable airlock (which the Soviet Union used on its first spacewalk in 1965), intended to facilitate Extravehicular Activities (EVAs) from one spacecraft to another.

But I've saved the best for last: there are indications that the air force intended to test the Gemini capsule for use as a space interceptor. Yes, that's just what you think. The euphemism was to rendezvous with "non-cooperative targets." Now, this could include a rescue mission to a disabled capsule, or a rendezvous with a spent rocket stage for research purposes . . . but more intriguingly, Soviet satellites and even manned spacecraft may have been on the docket. After all, these guys were pilots, and military pilots have military agendas. The orbital paths specified in the relevant Blue Gemini documents coincided with, among other things, the orbits used by Soviet spacecraft.

The planning continued, and at one point there was even talk of putting the entire Gemini program under the air force. Surprisingly, this suggestion did not come from the blue suiters, but from then defense secretary Robert McNamara.[6] Both NASA and the air force protested for their own reasons—NASA wanted Gemini to be oriented specifically for its Apollo-driven goals, and the air force still had aspirations for its winged spaceplane designs. From the air force's point of view, running the entire Gemini program would be far too demanding. In any case, Blue Gemini was eventually canceled, and the air force never got its man-in-space program.

But good ideas are hard to kill, and some of the air force's experiments and flight goals were absorbed into NASA's Gemini program. Twelve Gemini missions were flown, two unmanned tests and ten manned missions. Each flight had a specific set of goals that were met with various degrees of success, but all the missions flew generally as planned, and overall the program was a successful proving ground for the Apollo missions, and also met many air force goals. Some of the key accomplishments:

Gemini 3: The first manned flight of the new Gemini spacecraft.
Gemini 4: The first American "spacewalk" or EVA—Ed White opened his hatch and exited the Gemini spacecraft, drifting

free and testing a hand-held nitrogen-powered maneuvering unit for twenty-two minutes. A shared goal of both NASA and the air force.

Gemini 5: A week-long endurance test of two astronauts flying in a cramped cockpit, shoulder to shoulder, in increasingly uncomfortable space suits. A shared goal.

Gemini 6: See Gemini 7, below.

Gemini 7: Intended to test rendezvous techniques with an Agena rocket stage, an unmanned drone that had a restartable rocket engine that was designed to allow a docked Gemini capsule to maneuver from one orbit to another. The mission plan was changed at the last moment when the Agena blew up during launch. Instead, Gemini 7 became the target and waited in orbit until the quickly prepared Gemini 6A launched and eventually rendezvoused with it. They could not dock—there was no mechanism that would accommodate that—but they flew in formation for hours. Gemini 6 then returned to Earth, and Gemini 7 continued on another long-endurance flight—this time two weeks. These two missions also encompassed some of the air force's aims for Blue Gemini.

Gemini 8: The Gemini capsule successfully docked with an Agena stage, but problems developed quickly, and the flight was terminated early (see chapter 9). Another air force goal, docking with the Agena, was met nevertheless.

Gemini 9: Another rendezvous flight, during which another Agena exploded and had to be replaced at the last moment by a simpler target vehicle, which *also* suffered a malfunction when its nose cone did not eject properly. This flight was particularly interesting for a couple of reasons. First, it included another set of EVAs, but this time the spacewalking astronaut, Gene Cernan, would complete a set of work-simulation tasks. He was also supposed to climb to the rear of the Gemini capsule and don an AMU jet pack for testing in space. Unfortunately, Cernan became prematurely exhausted during the EVA—the difficulty of the procedure had been grossly underestimated— and the AMU flight was canceled after about ninety minutes of increasingly frustrating effort. It would have been another piece of the air force goals-in-spaceflight puzzle completed had it worked. As it turned out, free-flying with a rocket back- pack would have to wait another twenty years until accom- plished during the shuttle program in 1984.

Gemini 10: The Gemini spacecraft successfully docked with an Agena stage and was able to fire the Agena's rocket engine to change to a higher orbit—a critical test for rendezvous, docking, and (from the air force's perspective) the intercep- tion of other space vehicles. It was a success and another air force goal met. There was also another set of EVAs, and associ- ated tasks, that met with mixed results.

Gemini 11: Similar to Gemini 10, the Gemini capsule mated with an Agena booster and fired its engine to reach a record orbit, increasing from its original altitude of about 160 miles to a high of 740 miles. More EVAs were conducted with a bit more success.

Gemini 12: The final flight of the program, the Gemini spacecraft again rendezvoused and docked with an Agena. Buzz Aldrin conducted the final spacewalks of the Gemini program and hit his marks, completing the assigned (and simplified) tasks successfully, due to lessons previously learned in previous flights and arduous training on the ground before liftoff. Also of note, during the ren- dezvous attempt, the Gemini's navigation computer failed. Aldrin dug out a sextant, and between himself and pilot Jim Lovell (later of Apollo 13 fame), they navigated, rendezvoused, and docked manually, using visual sighting and manual hand controllers to control the maneuvering thrusters. This mission was a tour de force and met more of Blue Gemini's goals.

The air force officials and others involved with Blue Gemini's brief life (lived mostly on paper) likely felt mixed emotions about the Gemini program—their own space programs were sputtering (though another, the Manned Orbiting Laboratory, would still be pursued beyond the end of the Gemini program). Yet, many of the goals and techniques they had specified for their "blue" missions were met and mastered to one degree or another during NASA's flights. The data and experiences from Gemini were invaluable to the Apollo flights in 1967, and would have been very useful to other missions planned by the air force.

The idea of a manned space interceptor still intrigues. The image of a Gemini capsule, mated to an Agena stage for propulsion and rendezvousing with "non-cooperative targets," is compelling. When docked and under power, the Agena-Gemini combination was the first US manned spacecraft tested for an in-orbit restart of its engine and for performing major orbital altitude changes. Since the two craft docked nose to nose, it also had the disadvantage of forcing the astronauts to fly backward, heatshield first, when in use. But in theory it would have enabled extensive orbital maneuvers (limited, of course, to the Agena's fuel supply and by the reentry requirements for the Gemini crew), and the possible interception of Soviet spacecraft—robotic or manned. The purpose of this sort of mission could have ranged from close observation of the USSR's space activities to the disabling of a Russian spacecraft in flight. Either would have been incredibly provocative and risky, but were surely in the air force's planned bag of tricks. Successful use of the AMU rocket pack would have upped the ante further.

Blue Gemini would have been an interesting program for the air force, but ultimately most of what they hoped to accomplish was done just as well by robots. Other parts of the project, such as using the AMU or Gemini capsule to maneuver close in to a Soviet satellite or manned spacecraft, turned out to be unnecessary. In the end, the air force met most of its important goals in space with robotic spy satellites, and NASA flew ten manned Gemini flights and met its own goals on the march to a lunar landing. The right people did the right jobs after all.

CHAPTER 9

FLIRTING WITH DEATH: THE TERRIFYING FLIGHT OF GEMINI 8

UNCLASSIFIED

Neil Armstrong was in a cold sweat, using every bit of skill he had honed as a test pilot in some of the most dangerous planes ever designed. Dave Scott, sitting only inches away in Gemini 8's right-hand seat, was scanning the instrument panel furiously, looking for a clue . . . any clue. As Armstrong continued to struggle with the thruster handle in his left hand, firing this way, then that, he was simultaneously compiling a list of what he'd tried to arrest the motions of the wildly gyrating spacecraft. It just didn't make sense. Another part of his brain was less sanguine, because it was tracking what his eyes saw out the small, oval, eye-slit window just two feet in front of him: Earth, stars, and space were spinning past, at an increasing speed. Again and again, bright and dark, bright and dark, his vision starting to gray out at the edges as the spin increased in violence. Soon none of this would matter. The human body can only take so much, and even the most punishing turns in the simulator back in Houston had not prepared him for this . . . because it could have ultimately been deadly.

Gemini 8 was spinning out of control in space, rotating 360 degrees once every second, sixty revolutions per minute, and both men in the cockpit knew that they had only seconds to put a stop to it before they would black out and die. If that came to pass, Houston would try to raise them on the radio for days, to no avail. The spacecraft would just keep spinning for months in the vacuum of space, with two dead astronauts slowly decaying inside their suits. It could mean the end of the program, and nobody wanted to prevent that from happening more than the man at the controls. But he had no idea what to try next.

Fig. 9.1. The crew of Gemini 8: Commander Neil Armstrong, to left, and pilot David Scott, to right. Both would be tested to the limits of their experience as test pilots during the inflight emergency. Image from NASA.

By the time Gemini 8 flew in March 1966, the program seemed to have hit its stride. Gemini 3, the first manned flight after the tests of Gemini 1 and 2, had been routine. Gemini 4 had seen Ed White perform the first spacewalk in NASA's short history. Gemini 5 had flown for a week, setting a new US endurance record and using, for the first time, chemical fuel cells for power instead of batteries. Gemini 6 and 7 had performed an ad hoc rendezvous in space when their unmanned Agena docking target had exploded upon launch—the mission had been a masterpiece of on-the-spot planning and execution.

And now it was Gemini 8's turn to shine. This time the troublesome Agena rendezvous and docking spacecraft had launched successfully, and was silently orbiting the Earth awaiting the arrival of Armstrong and Scott to put it to use. The Agena was brimming with fuel and with its restartable engine—a first—would allow the capsule to dock and propel itself into a new orbit and practice various maneuvers. It was to be a critical test of technology and astronaut capabilities for the rendezvous and docking required for the Apollo lunar landings. Gemini was a "bridge" program, a series of utilitarian flights intended as a steep learning curve between the one-man Mercury missions, designed to get American astronauts into space, and the far more ambitious Apollo program to land humans on the moon. There were ten manned Gemini flights planned, and each one had a long list of objectives. This mission was no exception. It would be the first spacewalk, this time by Scott, since Ed White's, just the previous year. During the two-hour EVA, he would retrieve an experiment from the front of the Gemini capsule, then activate another on the hull of the Agena.

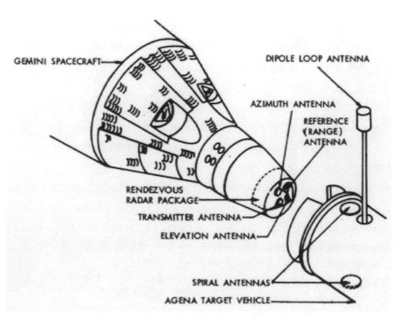

Fig. 9.2. Schematic of a Gemini spacecraft docking with an Agena. The conical docking collar on the Agena passively receives the nose of the Gemini, fastening docking clamps to hold it secure until undocking. Image from NASA.

This would also be the first docking of two vehicles in space—NASA planned to perform the maneuver four times—and the first change of orbit using a large rocket engine. The Agena even had its own set of maneuvering thrusters to allow it to be used to control the position of the twin spacecraft without depleting the Gemini's fuel. But the first task upon reaching orbit and getting settled in was to navigate to the Agena, dock, test its systems, verify that everything on both spacecraft—the Agena and the Gemini capsule—were in perfect order, then fire up the engine and hurtle into the record books.

But that's not quite how it went down.

After a routine launch and ascent, Armstrong and Scott settled into their preparatory routine, checking systems and getting ready to chase down the Agena already in orbit. Three firings of the Gemini's Orbital Attitude and Maneuvering System—the OAMS—changed their orbit enough to bring them within sight of the Agena, now about seventy-six miles distant. Armstrong switched the maneuvering system to the computer, an early model weighing over fifty pounds and positioned in the back of the spacecraft, to guide them to their target. About six hours after launch, they were at "station keeping," flying in formation with the Agena, during which they would inspect it for about thirty minutes prior to the first docking attempt.

Armstrong took control back from the computer and slowly edged the capsule toward the Agena's docking collar. Closer, closer . . . and with a thump they were docked. Latches snapped closed to make sure they stayed that way.

Armstrong said, somewhat laconically, "Flight, we are docked." Then, a few seconds later, Scott added, "It's really a smoothie."[1] Mission Control responded, "Roger. Hey, congratulations! This is real good." Armstrong added, "You couldn't have the thrill down there that we have up here."[2] And Jim Lovell, a fellow astronaut who was manning the CAPCOM or capsule communication console in Mission Control, just laughed.

A few minutes later, Lovell sent up a somewhat cryptic instruction: "If you run into trouble and the Attitude Control System in the

Agena goes wild, just send in command 400 to turn it off and take control with the spacecraft. . . ."[3] There had been some concerns about the systems onboard the Agena, so that was no surprise, but neither was it comforting. The Agena had been a troublesome vehicle since the beginning of the program. "Roger. We understand," Armstrong confirmed simply.[4]

And then, within minutes, the mission started to come unglued.

Scott first noticed that his attitude indicator, known to pilots as the "8-Ball" due to its spherical shape, was slowly moving off target. Something was not right—they should have just been drifting in space without changing direction. Armstrong gripped the controller handle and nudged the docked spacecraft back into position . . . and it started to drift again.[5]

Both astronauts assumed that the Agena was malfunctioning, understandable given their last-minute instructions from Mission Control. Scott sent the 400 command to the Agena, but the tumble continued. Armstrong was fighting it, but their maneuvering fuel was already down to 30 percent.

By now they were out of contact with Houston, flying over a dead spot between the ground stations tracking them, and would have to make a decision on the spot. Both spacecraft were now slowly spinning around their center of gravity, and if Armstrong waited too long to undock, the stresses of the tumble could jam the docking mechanism or cause the Agena to explode, killing them instantly. And they were low on thruster fuel, so time was critical. Scott agreed with Armstrong and then threw the switch to undock and to get away from the presumably malfunctioning Agena. But the drift continued.

As Armstrong maneuvered the capsule away from the Agena to avoid a possible collision, the spin picked up speed. It was puzzling—they had been certain that it was a problem with the other spacecraft. But clearly not—something was wrong with the Gemini maneuvering system. The capsule was their refuge, their ride home, and now that it was shorter and lighter after leaving the heavy Agena, the spin accelerated rapidly. They would have to think fast.

Soon they were back in contact with Houston via a navy ship stationed far below them for that purpose. Jim Fucci, the CAPCOM on the ship, knew that something had gone wrong. Clearly the flight was in trouble.

The first transmission Fucci heard was, "We have serious problems here. We're—we're tumbling end over end up here. We're disengaged from the Agena."[6]

Fucci responded, "Okay. We got your 'spacecraft free' indication here. . . . What seems to be the problem?"[7]

Scott said, "We're rolling up and we can't turn anything off. Continuously increasing in a left roll," to which the CAPCOM could only say, "Roger."[8] Clearly something grave had occurred. And Gemini 8's tumble was making data transmission spotty—the controllers on the ground had become spectators, and were working largely off the verbal transmissions of the increasingly stressed crew in space.

By this time the roll had become violent, with the sun flashing across the window every second; Scott later referred to it like being hit with a powerful strobe. It was disorienting, even to experienced test pilots like the two of them. And the intensity of the tumble was causing other problems—they were experiencing centrifugal forces more than three times normal. At about 4 g they would black out. Loose objects in the cabin were pasted against the walls, and it was becoming increasingly difficult to use the controls. At this point, each astronaut weighed the equivalent of nearly 600 pounds.

Armstrong continued to struggle with the controls, but it was not helping. Both now realized that they had a "wild" thruster—one of the maneuvering jets was firing out of control. They informed Mission Control of their status: "We have a violent left roll here at the present time and we can't turn the RCS's off, and we can't fire it, and we certainly have a roll . . . [a] stuck hand control."[9] And then, three minutes later, they were in another dead zone, out of contact.

There was one option left to try—the reentry control system, or RCS.[10] These were a set of small rockets separate from the orbital maneuvering thrusters, normally saved for controlling reentry. The problem was, once the RCS system had been activated and fired, even once, mission rules dictated that reentry follow swiftly. The valves could leak after a single opening and closing, and if they did, there would not be enough fuel to assure a controlled reentry—they could burn up. Using the RCS was a last-ditch option and would end the mission.

With no other choices left and his vision graying out, Armstrong disengaged the computer and fired the reentry control system. The sixteen thrusters began firing at once, and it took time—everything seemed to be taking forever—to finesse the capsule into compliance. But within moments, the spin was slowed, and Armstrong eventually regained control of the spacecraft. Gemini 8 was now drifting through space pointed forward as intended.

After thirteen minutes, they were in contact with the Hawaii tracking station. Scott radioed down: "Okay. We do have the Spacecraft under control at the present time. We're in slowly drifting flight . . . do have a slight amount of control."[11] They sent down their fuel readings, and the decision was immediate—come home.

But reentering and landing in the proper place is a complex procedure, requiring extensive calculations and preparation. Once new coordinates had been established for the proper location in the ocean—American space capsules were not intended to descend to dry land—naval ships would have to be positioned for recovery of the capsule and crew. Mission Control had to work fast, and the slide rules were out and calculating furiously.

The RCS fuel was still a concern—a lot of it had been used to regain control. Armstrong would still need a reserve to position and orient the spacecraft to set up their reentry. With only two degrees of latitude, the reentry angle was critical. And as they plunged into the atmosphere, the thrusters would also be working to keep the spacecraft in the proper orientation—the heatshield had to be pointed in the direction of reentry to protect them from the searing heat caused by friction with the air. There was precious little room for error now.

The CAPCOM read up long, numeric strings that Scott punched into the computer. After the adrenaline-pumping experience of just an hour ago, it was hard to settle down and go through the routine required for a safe ride home. But that is what training and simulation was for, and they had been through this part of the process hundreds of times. Regaining control of a wildly spinning capsule with a stubbornly stuck thruster, however, had not been closely modeled for simulation—that had been seat-of-the-pants flying.

Soon they were positioned to come home. The four powerful reentry rockets (separate from the RCS system) fired right on the money, burned for twenty-four seconds, and the astronauts began their thirty-seven-minute ride to the Pacific Ocean, 185 miles below. During the early part of reentry, communication with the ground would be increasingly spotty, ending in a long blackout period as they were enveloped in a cloud of ionized, 3,500°F heated atmosphere. All the controllers on the ground could do was wait, smoking furiously or tapping their pencils on their consoles, eyes fixed on frozen readouts.

Ten hours and forty-one minutes after their textbook launch, Armstrong and Scott were bobbing in a moderate Pacific swell, not far from their recovery ship. The destroyer *Leonard Mason* had driven at flank speed to reach the predicted splashdown zone, and despite everything, Gemini 8 was right on target. A short time later the astronauts became seasick and lost what little remained in their stomachs from the flight. Then the welcome sound of a navy diver rapping on the hatch signaled that their ordeal was over.

A postflight analysis indicated that a thruster had gotten stuck in the open position, firing continually and forcing them into an unrecoverable tumble, until Armstrong switched over to the reentry control system. The Agena had been fine.

Dave Scott would go on to command Apollo 15, and be the first man to drive on the moon in the lunar rover. Neil Armstrong, of course, gained fame as the commander of Apollo 11 and the first human to set foot on the lunar surface. But their substantial mettle was first tested in the hair-raising flight of Gemini 8, the ultimate challenge for test pilots in space.

CHAPTER 10

MANNED ORBITING LABORATORY: HOW TO DESIGN, TEST, AND NEVER FLY A SPACE PROGRAM

CLASSIFIED: *DECLASSIFIED IN 2015*

I t was to be the ultimate two-way stare down of steely-eyed Cold War missile men: while frustrated Soviet generals grumbled over radar images of a capitalist-spawned steel tube orbiting insolently above their country, Americans in that tube would be taking rapid-fire telescopic images looking back down at Soviet military installations and ICBM sites. The Russians would ponder their options for striking the orbiting spy outpost with missiles, and the astronauts inside the space station would feel the hair on the backs of their necks prickle a bit, even as they enjoyed their position of presumed impunity. This was the US Air Force's vision for the Manned Orbiting Laboratory, or MOL, intended for flight in 1968, but ultimately destined to be flown only in locked file cabinets in largely forgotten air force archives.[1]

The Manned Orbiting Laboratory and Blue Gemini military space programs fell under the following broad dictate issued by a task force assigned to evaluate the military uses of space:

"It is almost certain that as man's conquest of space proceeds, manned space stations with key military functions will assume strategic importance. It is therefore prudent for the Air Force to undertake R&D programs to explore the capabilities and limitations of man in space; to undertake exploratory development of special techniques to exercise military functions from manned orbital bases, and to program flight tests of primitive manned orbital bases with the capability of rudimentary military functions."[2] This study defined the Gemini spacecraft, already under development by NASA, as a key component of US Air Force plans to militarize space. The projected budget was $10 billion, or just under half of the eventual cost of the Apollo program. Experience from the separate Blue Gemini program would be critical to the success of MOL, since it would provide flight time for air force astronauts—and the astronaut's trip to and from orbit would be inside an air force Gemini B capsule. Rendezvous and docking experience with the spacecraft would be a bonus for more advanced versions of MOL. This also marked the end of the road for Dyna-Soar, which was a completely different type of spacecraft (a winged rocket plane without docking capability) and did not fit into the new, modular plan of MOL and the new and improved Gemini B spacecraft that would ride atop it.

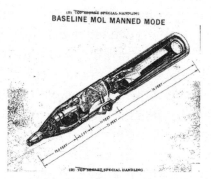

Fig. 10.1. A diagram of the Manned Orbiting Laboratory, which was classified for decades. The general shape of MOL was known, but the contents of the back half of the orbiting platform—top right—was not. It was to contain a giant spy telescope, intended to photograph the Soviet Union, among other regions the US wanted to monitor. Image from NRO archives.

Manned Orbiting Laboratory was a benign-sounding name for a space station intended primarily for spy duty, and it was the air force's last grasp at a role in manned spaceflight. Like Blue Gemini, the hardware was to be an outgrowth of the Gemini program, using the variant Gemini B capsule, mated to a long, slim orbital module not unlike

NASA's much larger Skylab of 1973. There was one big difference, however: while Skylab carried a telescope aimed at deep space, MOL would have carried one that would be aimed primarily at Soviet military installations. The top-secret program was publicly acknowledged as a research station in the 1960s (the Revell company, maker of hundreds of plastic models, even created one of MOL in the 1960s), but would not be fully declassified until 2015. The copious documents—some 20,000 pages—were released in print and posted on the Internet by the National Reconnaissance Office (NRO) and contained a few surprises. Among these insights was the painful and tortured path the MOL space station followed from its inception to its cancellation, one of the largest and most expensive space efforts to never even come close to flying. It is a study in shifting priorities, redundant and obsolete doctrine, and the displacement of military astronauts by robotic technology. In the span of six years, humans became largely unnecessary, at least for orbital spy missions. It was the beginning of a debate that continues today about the value of crewed civilian spaceflight.

The MOL program was born the same week the air force's X-20 Dyna-Soar spaceplane was canceled. It was a way for the air force to stay in the space game, at a lower cost, sharing technical development with NASA (the MOL space station would be overseen by the air force, as would the booster, but development of the Gemini B capsule would be shared with NASA and McDonnell Aircraft). MOL was announced to the public in 1963 by the air force, then again in 1965 by President Lyndon Johnson.

The 1963 announcement read in part, "AIR FORCE TO DEVELOP MANNED ORBITING LABORATORY: Secretary of Defense Robert S. McNamara today assigned to the Air Force a new program for the development of a near earth Manned Orbiting Laboratory (MOL). The MOL program, which will consist of an orbiting pressurized cylinder approximately the size of a small house trailer, will increase the Defense Department effort to determine military usefulness of man in space. This program, while increasing this effort, will permit savings of approximately $100 million over present 1964–1965 military space programmed expenditures. MOL will be designed so that astronauts can move about freely in it without a space suit and conduct observations and experiments in the laboratory over a period of up to a month. The first manned flight of the MOL is expected late in 1967 or early in 1968."[3]

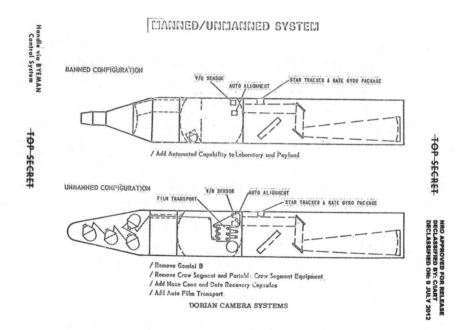

Fig. 10.2. At top is a schematic of the manned MOL mission, and at bottom an unmanned one. Again, it's clear in this illustration that the mission was all about reconnaissance—over half of MOL was devoted to the spy telescope. In the end, even the unmanned variant was not built due to advances in robotic satellite design. Image from NASA.

By the time of the 1965 announcement, the general design of the orbiting lab was made public through illustrations and cutaways, and flight crews were identified. The enormous top-secret telescopic spy camera was not—illustrations did not show any accurate detail inside the roughly 60 percent of the orbital platform that would contain the huge instrument. MOL was billed as a scientific research outpost with military capability, though the latter was only vaguely defined. Only with the recent release of details about the observing program, i.e. detailed surveillance of the Soviet Union, did the full extent of MOL's capabilities become clear.

On August 25, 1965, President Johnson announced the MOL program publicly, extolling its many peaceful virtues:

> I am today instructing the Department of Defense to immediately proceed with the development of a manned orbiting laboratory. This program will bring us new knowledge about what man is able to do in space. It will enable us to relate that ability to the defense of America. It will develop technology and equipment which will help advance manned and unmanned space flights. And it will make it possible to perform very new and rewarding experiments with that technology and equipment. . . . The Titan 3C booster will launch the laboratory into space and a modified version of the NASA Gemini capsule will be the vehicle in which the astronauts return to earth. . . . We believe the heavens belong to the people of every country. We are working and we will continue to work through the United States—our distinguished Ambassador, Mr. [Arthur] Goldberg is present with us this morning—to extend the rule of law into outer space. We intend to live up to our agreement not to orbit weapons of mass destruction, and we will continue to hold out to all nations, including the Soviet Union, the hand of cooperation in the exciting years of space exploration which lie ahead for all of us.[4]

This was all true in concept, but was a staggeringly incomplete description of MOL and did not address its primary mission goals: the militarization of space, development of highly accurate and powerful orbital spying, and ultimately the ability to control nuclear war—and its aftermath—from a safe haven in orbit.

In a not dissimilar vein, the air force tried to present MOL in a light that made sense to non-military minds. Robert McNamara, then secretary of defense, suggested that MOL would be a national space station: "[T]he Orbital Space Station Program [is] one requiring a new national mission to be assigned by the President on behalf of all national interests." The implication was that since NASA had been assigned the lunar landing mission, the air force could tackle more complex orbital efforts such as MOL. A statement by the secretary of the air force, Eugene Zuckert, added, "Both because of the primary commitment of the NASA to the manned lunar landing program and because of the important military interests in near earth orbit."[5]

Others shared these concerns. Dean Rusk, the secretary of state, wrote in a memo to McNamara that public conversations regarding the MOL program needed to have scientific "cover" to be acceptable to other nations. In a 1965 memo, he stated, "In terms of potential foreign reactions, I consider it most important that to the extent it [announcements of MOL's purpose] can be controlled, everything said publicly about the MOL project emphasize its experimental and research nature, and that statements and implications that MOL constitutes a new military operational capability in space, or an intermediate step toward such a capability, be rigorously avoided."[6]

MOL was elegantly straightforward in form. At fifty-seven feet in length and ten feet in diameter, it was smaller than NASA's later Skylab (which was eighty by thirty-three feet), but still a comfortable size for its crew—"approximately the size of a small house trailer," as the air force put it[7]—at about 400 cubic feet. However, unlike Skylab (which is still the largest single pressurized vehicle ever flown), under half of MOL would be a pressurized shirtsleeve environment—hence the seeming disparity between livable volume and the overall size. The rest would be given over to experiments and equipment—much of which was for spying purposes—and housing a giant, and extremely sophisticated, spy telescope. MOL would weigh about 32,000 pounds (Skylab was just over 80,000 pounds).

MOL was to be launched by an evolved version of the Titan ICBM that powered the Gemini program. This version, the Titan III, had huge strap-on solid rocket boosters, not unlike the space shuttle's SRBs. Atop MOL would ride a newly designed Gemini B capsule, carrying the crew aloft along with the station. To avoid the need for the crew to undertake an EVA to move from the capsule to the station, the Gemini B capsule would have a modified heatshield with a circular hatch leading into MOL.

MOL would launch as a single unit, shedding the Titan stages as it neared orbit. The Gemini capsule would stay affixed to the station during its entire mission of thirty days. When it was time to come home, the astronauts would transition back into the Gemini capsule through the heatshield hatch, detach from the MOL, and reenter as other Gemini flights did. MOL would then be placed into a controlled reentry to burn up in the atmosphere. In summary, it was a modest and workable approach to what would have been the world's first space station.

A crew of mostly air force pilots was recruited to fill the flight roster—among them a number of names that later became familiar to the public as NASA astronauts—once they had migrated to the space agency after the cancellation of MOL.

The core of MOL, and the part not disclosed until recently, was the massive camera that made up the bulk of the space station's length. Code-named "Dorian," the massive telescopic camera was developed by Eastman Kodak, which was also working on unmanned spy satellites for the air force. A version of this camera flew on the Corona and Gambit satellites, with Corona starting in 1959 and Gambit in 1964.[8] By the time of Gambit, the optics provided a resolution of four feet (later improved to two feet) using 70mm film. MOL would have done far better. Dorian was much larger, with a massive set of seventy-two-inch objective mirrors and a longer focal length, allowing for unprecedented image resolution.[9] It would be aimed via smaller "spotting" telescopes that would allow the MOL astronauts to get a low-magnification view of potential targets, and then the huge Dorian telescope would be targeted to the proper coordinates. This offered many advantages over the primitive robotic systems then in use—besides the larger size and greater abilities, targets could be chosen quickly with operators onboard the spacecraft, and they would be trained in on-orbit maintenance and repair, a luxury not afforded to unmanned satellites. Despite any efforts at science in microgravity, MOL would be, at its heart, a massive orbiting telescope, pointed not up at the heavens, but down at potential threats and targets.

Insofar as MOL would encompass other military capabilities, this was hinted at in another McNamara memo, one directed to then vice president Lyndon Johnson: "[T]here is the probability that it will evolve into a vehicle which is directly used for military purposes. It may provide a platform for very sophisticated observation and surveillance. Detailed study of ground targets and surveillance of space with a multiplicity of sensors may prove possible. Surveillance of ocean areas may aid our antisubmarine warfare capabilities. An orbital command and control station has some attractive features. While orbital bombardment does not appear to be an effective technique at the moment, new weapons now unknown may cause it to evolve into a useful strategic military tool as well as a political asset."[10] So, while the specter of orbiting nuclear weapons platforms had so far been avoided in the US, the hint of possible future weapons deployment was there. But clearly, even without this capability, McNamara was convinced of MOL's utility as a surveillance tool.

As the program evolved, challenges to its existence continued to come with every budget review cycle. NASA was forging ahead with its manned space program already, with or without air force efforts, and robotic systems were continuously improving. While early spy satellites such as Corona had been technically challenging (about a third of the satellites failed to return useful imagery), and Samos had been a failure (it *never* worked), the subsequent Gambit satellites returned an acceptable cache of images of Soviet (and Chinese) military installations.

One of the people brought in to evaluate the viability and scope of MOL, Dr. Harold Brown, who worked for the director of Defense Research and Engineering, described the human role for MOL cryptically (but honestly) before a US Senate committee: "If you just send a man up there without knowing what experiments he is going to do when he gets there, what you are likely to find is that everything he can do you have a machine that can do just as well. I am gradually becoming convinced that there are some things he can do better, but

I want the experiment specified first so when he goes up there he will actually be able to show he can do better. I think I can give you one specific example: I think a man can probably point a telescope more accurately than automatic equipment can. However, unless you design the equipment to measure that before you send a man up, and unless you give him a piece of equipment that will answer that question ... you are not going to get the answer."[11]

It sounds like an obvious statement, but it was really more incisive. There was a lot of interest in getting air force (and NASA) astronauts into space. Arguments for the utility and value of human operators could be manufactured by just about any organization involved in the effort—Brown wanted to make sure that the results warranted the expense, that humans could be more useful in some conditions than machines, and that the machines and activities scheduled for manned intervention should be designed in such a way, and with sufficient forethought, that the usefulness of human astronauts would be obvious.

Earlier in the life of the MOL program, a list of objectives for it had included:

- Acquisition and tracking of ground targets—spotting targets from a pre-assigned list.
- Direct viewing for ground and sea targets—scanning and spotting "targets of opportunity."
- Electronic signal detection—human-guided signal interception.
- In-space maintenance—the ability to repair military equipment in space.
- Extravehicular activity—to work and repair machinery outside the spacecraft.
- Remote Maneuvering Unit—test and evaluate remote-controlled craft.
- Autonomous spacecraft position fixing and navigation—the ability to navigate in space.
- Negation and damage assessment—observe and assess effects of military action.

- Multiband spectral observations—spotting missile launch and flight "signatures."
- General performance in military space operations—the measurement of human efficiency and psychological changes over the course of a MOL mission.
- Biomedical and physical evaluation—changes to the body during extended operations in zero-g.

Some of these were more relevant than others, but the list is remarkably similar to that established for NASA's long-duration orbital flights with the exception of specifically military tasks. The two-week flight of Gemini 7 would include some of these experiments, as would Apollo 7 and eventually the Skylab missions of the mid-1970s.

By the time the MOL program was moving ahead, the above list of "selling points" had been slimmed dramatically, and the purpose of the program was more exactly rationalized as a military reconnaissance effort.

From its inception, MOL was a program under pressure. The air force wanted to make sure that its scheduling stayed competitive with the Gemini and Apollo programs. Initially, the first unmanned MOL was supposed to perform a flight test in 1968, with a crewed launch in 1969. This was not acceptable to some members of senior leadership, and the project managers came back with a revised (and utterly impossible) schedule that shaved about two years off the first draft.

At the same time, NASA was contracting studies for space stations ranging from small two-man orbital research modules to twenty-man outposts. Most were based on existing and planned NASA hardware, primarily Apollo-based. There were concerns about overlapping efforts, and the air force suggested that MOL could be a worthy and useful forerunner of NASA's generally more ambitious efforts. It did not take long for NASA to blunt this idea. James Webb, the NASA administrator, stated that he felt that any "rigid" interpretations of such programs were not helpful, and that Gemini, Apollo, and MOL were all "important contributors to the ultimate justification and definition of a national space station."[12] While offering lukewarm support for MOL, Webb's statement offered little protection from cost-trimming politicians.

MOL soldiered on, receiving modest budgets for R&D while the Gemini and Apollo programs spent with comparative abandon. Air force studies were conducted to prove that human operators could accomplish meaningful tasks in a space-borne laboratory even while NASA flew astronauts in space. MOL hardware was prototyped and tested by the contractors while NASA tested its own spacecraft in orbit, which must have been increasingly frustrating for the MOL team.

Efforts were made to coordinate the upcoming Apollo flight hardware with MOL's objectives and designs, with NASA contracting studies of how to do so by their respective contractors. But now a new problem reared its head: while desirous of the study results, the air force could not release all the details (or, in some cases *any* details) of its own program goals for security reasons. NASA went ahead and studied the publicly available components of the mission but was not cleared to know about the "black" elements. In the end, these studies, like so much else that surrounded MOL, accomplished little.

And then, in 1965, another issue arose, a time bomb ticking away at the contractor for MOL. McDonnell Aircraft was completing the last Gemini program components, and the team that was so carefully assembled to eventually build the Gemini B capsules would soon be disbanded unless a new contract was approved. The expertise and capabilities of such a group were critical to MOL's success, and reassembling them later would be far too costly, if even possible. Decisions had to be made, and quickly—slips in MOL's schedule could kill the entire effort.

Meanwhile, the studies continued. They became more focused— requests for proposals (RFPs) were sent out, and a handful of aerospace firms bid on the construction of the MOL spacecraft. Other RFPs went out to Kodak and two other companies to begin substantive work on the Dorian telescope.

By late 1965, the MOL program was scheduled to begin flights in 1968. Six MOLs would fly—one unmanned test and five crewed flights through 1970. Real money was now being spent on design and fabrication, even as new concerns were raised. Once again, military planners must have been green with envy over the seemingly open pipeline of cash and support for NASA's manned spaceflight efforts. There was irony in this, since the military arguments for spaceflight were, in many ways, more compelling than those for a lunar landing. But President Kennedy had set his sights on the moon, and Apollo seemed inevitable. MOL, on the other hand, was now met with a resurgent concern about manned overflights of enemy territory. Interestingly, this concern centered on public reaction in the US as much as any rancor from the USSR. Additionally, concerns about human astronauts performing effectively in space continued, despite the success of the Gemini flights. These would not abate until Gemini 7's two-week marathon in December 1965.[13]

Meanwhile, the development of robotic spy satellites continued, and the rationale for sending men in orbit to spy on the USSR continued to diminish. While the air force tried to preserve MOL's original mission—that of manned surveillance with the rapid on-orbit decision-making capability offered by a crew—McNamara directed that an unmanned version of MOL be developed concurrently. General Schriever was opposed to anything that might undermine the manned version of the project, and commissioned yet more studies that indicated that the best results would be gained by flying MOL with a crew. Objectively, these results may have been serving the original question less than the air force's desired result. Despite the crew-centered studies, which were convincing enough to warrant continuation of both versions of MOL, the writing was on the wall: robotics were, for the purposes of orbital reconnaissance, catching up with manned capabilities. Automated systems may not have been as reliable or responsive, but they were good enough, less risky, and ultimately cheaper.

This divergence of concept, manned vs. unmanned, continued into 1966, with the air force continuing to advocate for the inclusion of crews and the Department of Defense eying the lower price of robotics. The different viewpoints might have had more to do with doctrine than anything else: the military wanted humans included in the great game of space dominance, and the civilian government wanted the most affordable system that could do the job (and to avoid political complications). The studies dragged on, slowing development of either version. Schriever continued to push for men on-station, with one of his leading consultants summarizing the air force's viewpoint nicely—"man is a neatly packaged system to do many tasks."[14]

Compellingly, the cost difference between the two systems was not staggering—the manned MOL program was estimated at $1.8 billion, the unmanned system at $1.5 billion.[15] The cost of the many studies and hours devoted to the skirmish may well have been a significant percentage of that cost differential in the final analysis.

Schriever retired from the air force in August 1966. With him went much of the top-level passion for military men in space—he had been a strong advocate for getting the air force into orbit since the beginning of the space age. In a parting memo, he wrote, "[T]he conduct of manned military missions in space will become indispensable to the defense of the nation in the future." Schriever continued, "The inception of the Manned Orbiting Laboratory Program has given us the opportunity to bring into sharper focus a broader appreciation of the potentials of military space by now encompassing the uniqueness, flexibility, and responsiveness of man." And he concluded, "As operational space functions become more complex and more sophisticated with time, the need for the development of truly effective manned systems emerges with increasing urgency. There is no true alternative for a manned system...."[16]

A month later, McNamara indicated that he agreed. But budget cuts soon beset the program, in large part due to the endless escalation of the Vietnam War. The launch schedule also slipped, partly due to budget shortfalls and also because Eastman Kodak was having trouble delivering the primary instrument—the giant telescope—on schedule. By the time the planning issues were worked out with Eastman Kodak, the first planned launch of a completed ready-to-go system had slid into 1970. And with program revisions, the budget had risen to over $2 billion, then $2.4.[17] To make things worse, the gradual realization by the media of what the MOL program really was—an orbiting spy platform—resulted in much rancor, with the press being harshly critical of the program. There was a sense that America had two space programs with overlapping goals, and that the associated costs were too high. The Apollo program had already become increasingly unpopular, and continuing MOL was simply too much to ask.

Ongoing problems with the program's budget and components in late 1967 pushed estimates for the first fully capable flights into 1971. It did not help that Apollo 7 would fly in 1968 on a Saturn IB rocket, making the Titan III seem redundant for manned flights. But the MOL missions could still be accomplished given enough support, and the program continued into 1968....

And in that year, the United States seemed to come unglued. In January 1968, North Korea seized a US ship at sea, the *Pueblo*, creating a grave international standoff. Then within weeks, North Vietnam unleashed the Tet Offensive, an unexpected and large-scale invasion of South Vietnamese strongpoints at a time when it was widely assumed that the US was slowly winning the war. Costs for all the military services escalated to meet the invasion. Finally, as a coda to a rotten year, Robert Kennedy and Martin Luther King were assassinated, and riots ignited a number of US cities.

The final blow to the project came with the incoming Nixon administration in late 1968, coupled with opposition to MOL from an unexpected source—the CIA. The spy agency felt that unmanned satellite photography could get them sufficient high-resolution imagery to make the necessary intelligence estimates, and that the difference in cost between MOL and the next generation of satellites made servicemen in space undesirable.[18]

In June 1969, Nixon dropped the axe in a speech, terminating the program under the guise of the duty of the government to spend tax money wisely, and added, "There is no more justification for wasting money on unnecessary military hardware than there is for wasting it on unwarranted social programs. . . ."[19]—a clear dig at Lyndon Johnson's efforts to build his Great Society. MOL died a quiet death, with the final shudders being large payments owed the contractors to shut down their parts of the program. To date, over $1.3 billion had been spent,[20] almost the entire original budget. To be fair to the air force and its accountants, a notable percentage of this had been burned on studies, delays, and person-hours spent in endless meetings.

MOL's mission would ultimately be fulfilled, at least in part, by a new satellite series called Hexagon, with nineteen of the giant machines flying successfully. The resolution of the cameras onboard was about twenty-three inches—better than the Gambit series it replaced, but not even close to the one-foot target resolution of MOL. The Hexagon satellite was ten by sixty feet—not much smaller than MOL, and weighed between 24,000 to 28,000 pounds . . . again, almost the same as the target weight for MOL. Hexagon carried between four and five film retrieval capsules like its predecessors, returning exposed film via fiery reentry to be snatched from the skies by waiting airplanes. Eastman Kodak did not build the telescopic camera—the contract was given instead to the American optics company PerkinElmer.[21]

The first of the Hexagon missions sent a film capsule hurtling into the deep Pacific when the parachute failed. Months later it was retrieved by a deep-diving US submarine—the Russians must not be allowed to capture the film—but was a sodden mess and returned no images.

Not much of MOL was ultimately fabricated in functional hardware. The Astronaut Maneuvering Unit—the AMU—was eventually sent into space for testing aboard Gemini 9, but was never actually activated due to Gene Cernan's fatigue during his spacewalk. A similar unit was eventually tested successfully by the space shuttle in 1984.

Of the seventeen pilots selected as astronauts for MOL, seven ended up at NASA, including shuttle pilot and future NASA administrator Dick Truly, and shuttle pilots Robert Crippen, Gordon Fullerton, Hank Hartsfield, and Robert Overmyer.

The optical technologies researched and developed for MOL's giant orbital telescope eventually found their way into unmanned reconnaissance satellites, the KH-11 Kennan series, returning highly detailed images of the Soviet Union and China that were reported to have eventually achieved imaging resolution of three inches and a new capability—video transmissions. Observers from the ground in the US could have watched Soviet generals making a vodka toast from the vantage point in space—just as MOL would have eventually been able to do.

MOL would have provided experience with humans in space just a couple of years before Skylab did, and in a much smaller, less scientifically capable package, so the loss to the US space program does not seem severe in hindsight. However, in its most advanced iterations, MOL would have provided an open-ended orbital architecture, a true infrastructure in space that was not achieved until the International Space Station flew. One suggested configuration of MOL had three of the stations joined to a central hub, and it could have serviced multiple Gemini spacecraft and allowed for continuous habitation.[22]

And there were other, alternative plans for MOL, beyond an observing platform. A 1965 memo titled "Advanced MOL Planning"[23] lists a number of intriguing possible missions, including "Strategic Warning," "Tactical War," "Precise Targeting," and "Command Post." The most advanced versions of this would weigh almost half a million pounds and support forty men—though it is not entirely clear what most of them would actually do. The multiple-MOL station is, however, listed as a "Command Post." There are a lot of redactions in the MOL-related materials released in 2015 and later, blacking out sentences and entire paragraphs, so it may be some time before we know the full capabilities of the proposed program in its entirety.

The Cold War was one fought mostly without bullets (certainly the US and USSR never fired on one another directly, but their Third World proxies did), and many of the planned space programs were "flown" only on paper. The Soviet Union would eventually orbit its own military space stations, code-named Almaz, in about the same time frame as MOL would have ultimately operated, starting in 1973. The public name for these stations was Salyut; four were built, and two of them were successfully flown. But the Soviet leadership came to the same conclusions that the US Air Force did, and eventually turned to the use of automatic satellites to survey its arch enemy. With this, the short career of manned military spaceflight came to a quiet close.

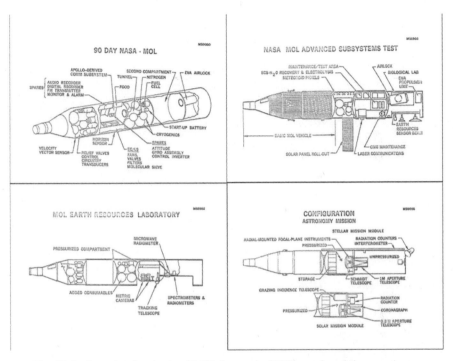

Fig. 10.3. A study of variants of MOL for use by NASA produced these variations on the spy-platform theme. Lower right, the "Astronomy Mission," is of particular interest due to the fact that it swaps out an astronomical telescope for the downward-looking spy unit. Image from NASA.

CHAPTER 11

APOLLO 11: DANGER ON THE MOON

UNCLASSIFIED

When July comes to Wapakoneta, Ohio, it feels hot enough to make the devil seek shelter. It's not the temperature strictly speaking, but the humidity—summer in the small town can feel like walking through hot, soaked cotton. Mornings often begin with a misleadingly temperate spell, and then the air gets moist and sticky and the world seems to shrivel. But on July 20, 1969, even the most delicate souls in the small Midwestern town did not mind one bit. One of Wapakoneta's own was about to do something amazing.

About 240,000 miles away, that son of Ohio, Neil Armstrong, along with his partner Buzz Aldrin, had made a weightless crawl through a short tunnel connecting the Apollo 11 capsule to the waiting lunar module, leaving the third member of the crew, Michael Collins, to a lonely vigil. He would orbit the moon alone as his two comrades gambled their lives on $32 million of technology, the cost of one of NASA's lunar modules, which had, as always, been built by the lowest bidder.[1]

Soon the two astronauts would disconnect from the command module, make only the second descent toward the moon's surface ever (Apollo 10 had performed a "dress rehearsal" in May, descending and coming back but not landing), and attempt to land. Collins wished them well, but in private thought their chances of success were about fifty-fifty.[2] They did in fact make a successful landing, but the entire scenario was ultimately jeopardized by a slug of ice in a fuel line that wouldn't have lasted ten seconds on an overheated Wapakoneta sidewalk.

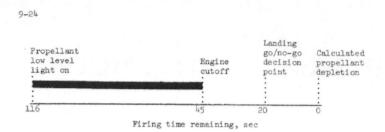

9-24

Fig. 11.1. Excerpt from the Apollo 11 lunar landing flight plan: the fuel level versus remaining time plot. The abort prompt is supposed to be triggered when the instrumentation senses twenty seconds of fuel left in the tanks. Armstrong and Aldrin were just past that time when they set down on the moon. Image from NASA.

Across the nation, people who had televisions sat, staring numbly at the screen—125 million of them,[3] or half the living souls in the country. Those without TVs listened to stunned reporters on the radio. Across the world, people stood still, and well over a billion bore witness to the incredible events happening across the void. As the moon walk became real within their minds, folks were nodding or shaking their heads in amazement or disbelief. While the actions varied, the emotions were the same. *We did it. We goddamn did it.* Fathers spoke to their children in weighty tones about this momentous moment in history, a day they would never forget, and waxed philosophical about the new world that lay before them. The children's wide eyes went from their normally stoic father's smiling face, then to the flickering black-and-white TV with the ghostly snowmen moving across a bleached landscape, trying to understand why Dad had that funny look in his eyes. Mothers told the kids to adjust the rabbit ear antennas. And while they did, some of those fathers crept outside into the backyard, took a sip of whiskey, looked at the moon, and wept. It was America's finest hour.

On the TV screen inside, the tense voices of Mission Control and the scratchy audio from the crew in the lunar module could be heard going through the landing procedures. The TV news anchors commented intermittently, confused by the unexpected computer alarm codes being relayed to Houston by Aldrin. They only interjected when something important had to be explained.

Apollo 11 was four minutes into its landing sequence when the terse words of its commander came from the speaker in Mission Control: "Program alarm."[4] Aldrin, standing next to Armstrong in the descending lunar module, stared at the frozen display on the computer, which read "1202." It was an error code, but for what? Controllers in Houston scanned their notes trying to figure out what the heck the problem was. But time was running short.

Fig. 11.2. Tom Kelly, in charge of the lunar module effort at Grumman, sits to left (facing camera) at the Manned Spaceflight Center in Houston. He and his engineer colleagues would sweat out the brief but scary pressure buildup in a LM descent stage fuel line, in constant consultation with other "Grummies" back at the plant in Bethpage, Long Island, and engineers at TRW, builders of the descent engine. Image from NASA.

"Give us a reading on that program alarm," Armstrong said. He sounded intense, but no more so than during the simulations. It was hard for the larger audience to grasp that a life-or-death struggle was playing out 240,000 miles (386,000 kilometers) from Earth, in a small, fragile machine descending rapidly to the moon. Communications were spotty; the computer was threatening to quit, and Gene Kranz, the flight director for this first lunar landing, felt Mission Control slip a bit further behind the power curve. All they could do was listen, and wait.

Most people knew that going to the moon was risky. Few outside of Mission Control, listening to the communication between the astronauts and Houston, understood the urgency of the message. Now the dangers were no longer theoretical; they were being played out in space at that very moment.

Much later, Gene Kranz related the stakes as he had outlined them to the controllers just before the landing. "We would either land on the moon, we would crash attempting to land, or we would abort," he said simply. "The final two outcomes were not good."[5]

The problems began immediately upon separation from the command module in which Armstrong, Aldrin, and Michael Collins had flown to the moon. Mission Control was having trouble with the radio link to the lunar module. "It was purely my decision how much information was enough," Kranz recalls. "This is now going through my mind: 'Do I have enough information to continue?' And the answer is yes"—but barely. "I gave the crew the go for powered descent"—that is, to begin the rocket-braked landing—"and we immediately lost communications again."

Aldrin had adjusted the antenna, and Mission Control had done what it could on its end, but the radio connection just kept on fading in and out. If it got much worse, Kranz would have to order an abort.

As the astronauts struggled with communication issues and spotty radar data, the master alarm sounded in the LM's cabin, also lighting console warnings in Mission Control. The landing computer was signaling an overload; the 1202 alarm it displayed was an error code that meant, in effect, "I have too much to do, so I am going to stop, reboot, and start over." This could have forced an abort of the landing. The computer was receiving more data than it could handle.

The events startled Aldrin; the code was unfamiliar to him. "We couldn't look it up in the book to see what the problem was because we were watching where we were going!" he later said.[6] Armstrong waited a bit, then asked Houston for clarification.

On the ground, controller Steve Bales finally made the call: it was okay to continue, so long as the alarm was intermittent. "In the middle of landing, it was almost as dangerous to try to abort with a bad computer as it was to carry on with the landing," Bales said. "So balancing risk versus risk, we decided that the safest thing would be to continue to land."[7]

CAPCOM Charlie Duke, a fellow astronaut, relayed the message to space: "We're go on that alarm!" Then, within minutes, the computer displayed another warning: 1201. But Bales indicated that it was the same class of warning, so they continued. "Same type, we're go," Duke radioed.

At this point, Aldrin was concentrating on the instrument displays, calling off the numbers for altitude, speed, and other critical data as Armstrong took over manual control of the landing.

But there was another problem—they were not where they were supposed to be.

Aldrin recalled: "In the commander's window was a grid, a vertical line with marks on it. And this was calibrated to his height and his eye level."[8] Taking cues from the computer, Armstrong could check against this grid to determine the LM's position over the moon. Now, it told him that they were coming in "long," or downrange. Between the lumpy gravity of the moon and some extra speed picked up when they undocked from the command module, Armstrong and Aldrin had overshot the predicted landing zone.

Armstrong stopped his descent as an unwelcoming view glared back at him from outside. Where a smooth plain was supposed to be, there was instead a vast crater field and collections of truck-sized boulders. Flying manually and low on fuel, Armstrong leveled off and searched for a smooth spot to set down.

Aldrin stated laconically, "You're pegged on horizontal velocity. . . ."[9] They were now skimming the surface as Aldrin kept his eyes glued to the instruments, monitoring fuel consumption, horizontal speed, and altitude. Armstrong just wanted something relatively flat to land on.

"I was looking at my trajectory plot," Charlie Duke remembered, "[and] Neil leveled off at about 400 feet [122 meters] and was whizzing across the surface. . . . It was far from what we had trained for and seen in the simulations. So I started getting a little nervous, and they weren't telling us what was wrong. It was just that they were flying this strange trajectory."[10]

In fact, they were flying for their lives. With a rapidly diminishing fuel supply—they would soon reach the sixty-second mark, when only one minute of fuel remained until a mandatory abort. And it was uncertain whether or not an abort—a forced dropping of the descent stage and emergency ignition of the ascent engine—would even be possible at this altitude. They were in a region called the Dead Man's Curve, and at a height at which the time required to stage and ignite the ascent engine might result in impacting the surface.

"You never [want to] go under the 'Dead Man's Curve,'" said controller Bales. "It was an altitude [where] you just don't have enough time to do an abort before you had crashed. . . . Essentially, you're a dead man."[11]

CAPCOM Duke called the sixty-second fuel warning. Armstrong focused intently on a smooth spot ahead. Aldrin continued calling out the speed and range.

"We heard the call of 60 seconds, and a low-level light came on. That, I'm sure, caused concern in the control center," Aldrin recalled with a smirk. "They probably normally expected us to land with about two minutes of fuel left. And here we were, still a hundred feet [30 m] above the surface, at 60 seconds."[12] Concern is a polite word for it. Mission Control was silent except for Aldrin's transmissions and the occasional sotto voce remarks of controllers monitoring their systems. Even CAPCOM Duke had fallen silent.

"Then there's another call for 30 seconds," Kranz recalled, referring to the ever-dwindling fuel supply in the LM. "And at about that time, you know, you really start to suck air, and I'm seriously thinking, now we have this land-abort decision that the crew is faced with. Are they going to have sufficient fuel to land on the moon, or are they going to have to abort very close to the lunar surface?"[13]

"When it got down to 30 seconds, we were about 10 feet [3 m] or less [from the surface]," Aldrin said. "I could sneak a look out, because at that point, I don't think Neil cared what the numbers were. He was looking at the outside. I could see a shadow of the sun . . . behind us."[14] In fact, the descent engine was kicking up so much dust at this altitude that the shadow and a few boulders sticking up through the haze were all Armstrong could use to gauge the remaining distance to the surface.

Then a long, metal rod that extended from the landing legs touched the lunar plain, signaling their arrival. A blue light on the console came on—"CONTACT LIGHT"—and the landing was over.

"Houston, Tranquility Base here . . . the *Eagle* has landed," Armstrong said calmly.

In Mission Control, there was a moment of continued silence. "It took us a couple seconds to really realize that we had made it. And then the people in the viewing room start stomping, and I mean just cheering and clapping and stomping their feet," Kranz remembered. "I'm so tied up emotionally at this time that I literally cannot speak, and I've got to get my team back on track." It took a physical act to get him back to reality. "I rapped my arm on the console—I broke my pencil!—and finally get back on track and call my controllers to attention. I say, 'OK, all you flight controllers, settle down. OK. Let's get back on with it.'"[15]

But even though the crew had successfully touched down, the moon had more mischief in store. Armstrong and Aldrin "safed" the LM, shutting down the landing systems and performing their post-landing checklist. Everything looked good on Mission Control's consoles. But within minutes, the controller monitoring the lunar module's descent stage signaled a potentially dangerous pressure buildup in a descent-engine fuel line.

Dick Dunne, Grumman's PR man for Apollo (Grumman built the LM), recalled the critical moment. The extreme cold from the lunar surface was creeping into the descent stage after engine shutdown. "The cold permeated a fuel line, and caused a blockage . . . which was immediately reported back by telemetry to Mission Control in Houston. That gave us cause for alarm," he remembered.[16]

The frozen blockage might melt, or it might cause a relief disk to blow and relieve the pressure. Or it might cause a catastrophic explosion. Nobody could be sure.

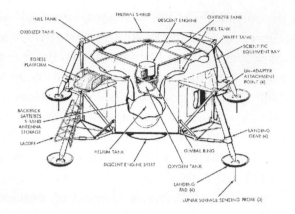

Fig. 11.3. Schematic view of the lunar module's descent stage. The ice plug that threatened to terminate the moon walk attempt was in a fuel line, near the center of the image. Image from NASA.

As Kranz debated whether or not to tell the astronauts, there was a quick conference among the controllers and the Grumman representatives. "There was some thought given to aborting the exploration of the moon and to initiate the launch sequence right away," Dunne recalled. "However, the heat that came out of the engine melted the ice that had formed, and the problem went away." Residual heat radiating from the red-hot engine crept up the fuel line, and pressures returned to normal. Controllers breathed a heavy sigh of relief, even as the astronauts continued their checklists, unaware.

Within three hours, Armstrong and Aldrin were ready to explore the moon. But as they prepared to exit the lunar module, yet another issue cropped up: they could not get all the air out of the LM. The astronauts had opened the relief valve and watched the pressure gauge as the oxygen vented out . . . but even though it read zero, they could not get the hatch open. There was still too much residual pressure inside the lander.

"We tried to pull the door open, and it wouldn't come open," Aldrin said. "We thought, 'Well, I wonder if we're going to get out or not?' It took an abnormal time for it to finally get to a point where we felt we could pull on a fairly flimsy door."[17] In fact, Aldrin eventually resorted to tugging on one edge of the front hatch. It was so thin that there was a bit of flex, and that was enough to vent the rest of the air into the lunar void.

Armstrong, aided by Aldrin, maneuvered toward the open hatch. As Armstrong twisted his bulky suit to head out, there was a tiny tug on his suit as something snapped on the instrument panel. Unheard in the vacuum of the cabin, Armstrong's backpack had broken off a breaker switch that would later be needed to arm the LM's ascent engine. Engineers in Houston conferred to come up with a work-around, but upon preparations to leave the moon some twenty-one hours later, Aldrin would calmly flip the broken stub with the point of a felt-tip pen. Problem solved.

The two astronauts departed the LM and explored the surface of the moon for over two hours. During that time they deployed numerous experiments, collected almost fifty pounds of rocks and soil samples, and took a congratulatory call from President Nixon.

On July 21, the LM's ascent engine ignited and carried Armstrong and Aldrin back up to a rendezvous with Michael Collins in the command module. With the entire crew docked and reunited, they depended on one more critical event to leave lunar orbit and head home. The rocket engine at the rear of the command/service module, the service propulsion system (SPS), had to ignite to break them free of lunar orbit. It was, like the ascent engine, a one-shot deal—no second chances.

"That service propulsion system has got to work," Kranz said. "Single option, big engine, that's the only thing that's going to get you home."[18] With the LM used up and discarded, there was no backup. The SPS firing would occur when the astronauts were on the far side of the moon and out of contact with Houston, so Mission Control would not know if the maneuver had worked until the capsule cleared the limb of the moon. Some controllers would later remember those last minutes of radio silence as the mission's longest.

But the SPS engine fired right on the dot, and Apollo 11 left the clutches of lunar gravity and sped home to a fiery reentry on July 24. The angle of entry into Earth's atmosphere was critical; at 25,000 mph (40,200 km/h), there was not much room for error. If the entry angle was off by even a couple of degrees, it could cause overheating and excessive, crushing g-forces, a vast overshoot of their landing zone, or even, in a worst-case scenario, a bounce off the atmosphere that could delay reentry until after the crew had run out of oxygen.

Kranz recalled those final moments of the mission: "It's a difficult time, a lonely time for the controller, because there's only one thought in every controller's mind: 'Did I get them to do everything we needed to? Is my data right?'" He grimaced at the memory. "No more give-backs. . . . We've trained well. They have confidence in us. We have confidence in them. And then it's up to them and the spacecraft to finish off the mission."[19]

Apollo 11's mission ended in the tropics of the Pacific Ocean, splashing down near Wake Island. In Mission Control, Gene Kranz and his team celebrated by clapping, whistling, and passing out cigars.[20] The first lunar landing was complete, and five more crews would follow to the moon's surface, conducting increasingly ambitious, and still highly risky, missions. But rising above it all, in Kranz's memory, is the overwhelming nature of the accomplishment. As he put it, "What America will dare, America can do."[21]

CHAPTER 12

THE FIRST SPACE SHUTTLE: PROJECT DYNA-SOAR

CLASSIFIED: *DECLASSIFIED IN 2011*

G ermany's *Silverbird* skip bomber may have never made it off the drafting tables of the Third Reich, but the designs for this visionary rocket bomber did not simply fade into the postwar haze of fantastic and unworkable wonder weapons. The US Air Force, ever on the lookout for newer and better ways to deliver nuclear destruction to the USSR, studied Sanger's plans closely (as did the Soviets). While there were problems with this early design, especially with reentry heating and the mission profile, the idea of a rocket-powered high-altitude bomber held appeal to the military mind. An additional enticement was that this was a winged craft that flew into space, under the control of pilots; this was sweet music to many within the air force establishment. And when viewed from their perspective in the early to mid-1950s, it made sense for the branch of the US military responsible for holding the high ground of the air to fulfill a similar role in space. Thus would begin the air force's largely futile fifteen-year quest for a role in manned spaceflight.[1]

This goal manifested itself in a number of studies commissioned by the air force through the 1950s, with the first being awarded in 1954 to Bell Aircraft. Bell had been the designer and manufacturer of the X-1, America's first rocket plane (in which Chuck Yeager first "broke the sound barrier"), so the company had a background in both rocket-propelled aircraft and hypersonic flight, and seemed a natural choice to build a space-borne version of a rocket plane.

Bell had already been working on plans for such a rocket plane when the air force engaged them. When Wernher von Braun brought his cadre of German rocketeers to the US after WWII, among them were Walter Dornberger, the military director of the V-2 program at Peenemunde, and Krafft Ehricke, another top-ranking V-2 official. Both these men ended up at Bell Aircraft by 1952, and many of the ideas, and much of the expertise, of the German plans for space came with them. This included an understanding of the Sanger skip bomber design.

This air force study by Bell represented the underpinnings of what would become the X-20 Dyna-Soar, the air force's first serious and well-funded (in relative terms) contender for a manned presence in space. In general, the air force wanted to realize the potential of the "antipodal bomber" or skip bomber as envisioned by Sanger with the *Silverbird*. There were updates to the idea, among them that the new craft would be launched via a rocket booster, either manned or unmanned, as opposed to Sanger's rocket sled. But otherwise the concept was similar.

The US Air Force contract was intended to allow Bell to continue work on a program they called BoMi, short for Bomber Missile. It was based on the *Silverbird* but attempted to resolve some of the shortcomings of that work, including, critically, reentry cooling.

The rocket plane that Bell had been working on when the contract was awarded was a two-part system, with a piloted booster stage and a skip bomber, also piloted. The booster would be 120 by 60 feet and the glider 60 by 35 feet.[2] The larger booster would take off from a runway with the bomber on its back, release it at altitude, and then fly back for a runway landing, refurbishment, and reuse. To cool the spacecraft during reentry, water would be circulated between the hot outer skin and the cooler inner aluminum airframe, with the hottest of it vaporizing and being vented out the rear of the craft. The glider would be capable of delivering a single 4,000-pound bomb across a range of 3,300 miles of suborbital flight.

When the air force hired Bell to complete a revised study, they changed the ground rules substantially. The range of the skip bomber would have to be increased from 3,300 miles to 12,000 miles. The required speed of flight was increased from 2,650 mph to *15,000* mph.

They also expressed concerns about the reentry cooling system and wanted more work done to prove its viability, especially in light of the grossly increased requirements.

A review one year later, at the conclusion of the contract, found that the skip-gliding concept was going to be more challenging than thought, even with Bell's revised figures and designs. The air force requirements meant that the craft had to boost into the upper atmosphere, then dive, slamming into denser air, and bounce off to generate lift for another bounce downrange. This technique simply generated too much heat and severe g-forces. So another flight technique was recommended: boost-gliding. Rather than skipping across the top of the atmosphere as Sanger had suggested, the rocket plane would fly into space, and then reenter in a long controlled glide. This would result in reduced heating and g-forces, while still providing acceptable range and flexibility. Most importantly, its preferred trajectory was perfect for long-range reconnaissance and bombing.

Another set of studies was requested from a half dozen aerospace contractors for a program called—almost cutely—RoBo, or Rocket Bomber. This was a similar boost-glide program, with a maximum altitude of 260,000 feet and a global glide capability at 100,000 feet. By 1957, the space-bomber concept had been adjusted and fiddled with a number of times, with the studies just rolling along . . . and then came Sputnik. When the Soviet Union lofted its beeping silver sphere in 1957, the effect was electric, and the US military took notice, to put it mildly. Of course, it bothered the air force to see its country upstaged . . . but not as much as seeing the Russians pulling ahead of its own efforts. By the end of that year, BoMi, RoBo, and yet another set of studies called HYWARDS were consolidated into what would become Dyna-Soar. This spacecraft would later be designated as the X-20, and it was perceived as the next logical step after the X-15. The X-15 was a high-altitude (but not orbital) rocket plane that began flying in 1959. What the X-15 would be for high-altitude, high-speed research, Dyna-Soar would be for space. And, as a huge bonus, it would not be simply a research craft, but tactical—a true orbital reconnaissance and weapons platform.

It was at this time that the American space effort was cleaved into two parts. The Eisenhower administration formed NASA as a civilian space agency in July 1958, and the onus of manned spaceflight was moved to the new agency, along with most of the available budget. The air force was stunned, but moved ahead with what funds it was able to procure. NASA would pursue the design of blunt-body ballistic space capsules, and the air force would fly into space and back in winged gliders. At least, that was the plan.

The specifications for Dyna-Soar were split into three sets of components to accommodate the overall goals of the program:

- Capable of a maximum altitude of at least sixty-two miles (one accepted definition of the boundary of space)[3]
- Able to provide a photo-reconnaissance, radar-scanning, and bombing capability
- Able to fly a range of 5,750 miles

By 1959, a partnership of the Boeing Company and Chance Vought Aircraft had won the bid to design Dyna-Soar. Martin (known then as the Glenn L. Martin Company) was given a separate contract to upgrade the Titan ICBM to a man-rated system. The Titan would also need to have higher thrust to carry what would ultimately become (on paper at least) a 16,300-pound spaceplane, 22,300 pounds with fuel and a booster stage. While this pales in contrast to the later space shuttle, which weighed about 240,000 pounds at liftoff,[4] it was still the largest spacecraft considered on the drawing boards at the time (except for the nuclear Project Orion, which was an entirely different program, and never prototyped—see chapter 4). A total of ten Dyna-Soar gliders were expected from the contract between 1965–1967.

The Dyna-Soar project would burn fast and bright. In its brief life, it would provide research applicable to the space shuttle well over a decade later. But Dyna-Soar itself would spend four tortured years working toward the prototyping stage before succumbing to political pressures and being canceled.

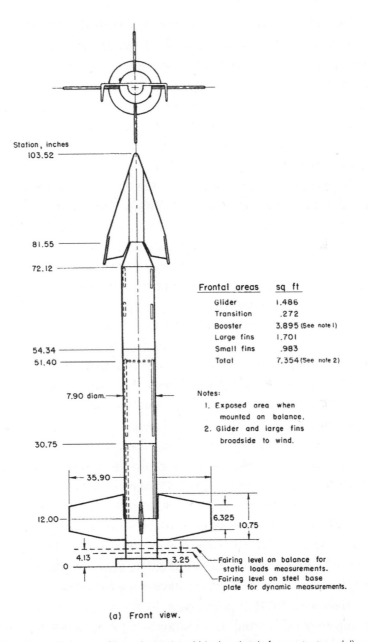

Station, inches
103.52

81.55

72.12

Frontal areas   sq ft

Glider	1.486
Transition	.272
Booster	3.895 (See note 1)
Large fins	1.701
Small fins	.983
Total	7.354 (See note 2)

54.34
51.40

7.90 diam.

Notes:
1. Exposed area when
   mounted on balance.
2. Glider and large fins
   broadside to wind.

30.75

35.90

12.00

6.325
10.75

4.13   3.25
0

Fairing level on balance for
static loads measurements.
Fairing level on steel base
plate for dynamic measurements.

(a) Front view.

Fig. 12.1. Dyna-Soar in profile and top view (this drawing is from a test model).
By placing the spaceplane on top of the rocket, instead of on the side as the later
space shuttle was, the pilot had an option to abort a malfunctioning booster by
igniting the top stage and pulling away from the Titan. Image from DOD/USAF.

The mock-up presented to the air force in 1961 was thirty-five feet long with a wingspan of twenty feet. It was the result of some 13,000 hours of wind tunnel testing and countless design revisions to address problems in the hypersonic speed range, just then becoming better understood via test flights of the X-15.

Special attention had been given to reentry. Rather than resist heat, as space capsules do with their ablative heatshields and the shuttle did with its protective tiles, Dyna-Soar would absorb it into the frame and dissipate it. The airframe and some of the upper body surfaces of the spaceplane were to be made from the same alloy as the Mercury space capsule, then under construction by McDonnell Aircraft, called René 41. This metal was not unlike that used in the X-15, in that it had a melting point close to what was expected for the average temperature of reentry in some areas of the spaceplane that were exposed to less heat. Dyna-Soar, like its smaller predecessor, would be operating at the edge of its material strength. Other exotic materials, such as molybdenum and zirconium, were used to protect high-heat areas. The exterior skin was separated from the airframe with insulation, and the outer hull plates overlapped to allow for expansion during heating. It was an ingenious design that would later be utilized by the Soviet Union for its own experimental spaceplane designs. The entire spacecraft was coated to prevent oxidization of the skin, which could reduce its ability to withstand the rigors of reentry. Each Dyna-Soar spaceplane was expected to go into space four times.

Windows were a special concern, and these were to be the largest pieces of pressurized glass ever flown—nothing this large actually went into space until the space shuttle flew in 1981. Unlike the shuttle, they would be covered with metal shields that would not be ejected until after the worst of reentry, and viewing would be entirely through side windows until that time. Dyna-Soar's design would have pushed it to the edge of existing technology. It is likely that the program, for all its bravado, would have been dogged and delayed by technical and developmental problems had it proceeded to fruition.

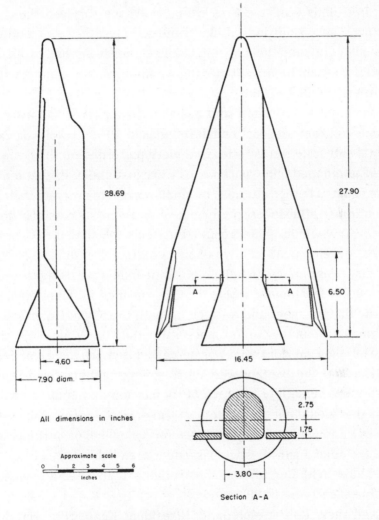

28.69

27.90

6.50

A   A

16.45

4.60

7.90 diam.

All dimensions in inches

2.75

1.75

3.80

Approximate scale

0   1   2   3   4   5   6

Inches

Section A-A

(d) Sketch of glider and transition.

Fig. 12.2. Dyna-Soar in profile and top view (this drawing is from a test model). The spaceplane was about thirty-five feet in overall length, with a wingspan of about twenty feet. While it was miniscule compared to the later space shuttle, if successful it could have given the US an entirely different type of access to space. Image from DOD/USAF.

The high-speed runway landing would have been accomplished on three skids—inflated tires would not have survived the reentry heating. These skids looked like truncated skis; the front skid was a solid alloy tray, and the rear skids looked similar but had exotic metal bristles on them to prevent melting when they made contact with the runway at about 230 mph.

To test these concepts and designs, the air force had awarded a further contract with McDonnell Douglas in 1961 to build six experimental sub-scale test vehicles that were patterned on the Dyna-Soar. These unmanned mini-spaceplanes were just under six feet long and were boosted to high altitudes by small rockets to evaluate their high-speed, upper-atmosphere performance, and were successful in demonstrating that winged gliders could function safely in that environment.

As development of Dyna-Soar continued into 1962, NASA's Mercury program was in full bloom. Lone men were being tossed into space every few months inside the tiny capsules, fulfilling the need to get the US into space and to catch up with the Soviet Union's accomplishments. Soon the Gemini spacecraft would be flying. While the air force felt the heat from what was perceived by many to be competition to Dyna-Soar, the overlaps were illusory. For one thing, Dyna-Soar had much more versatility in terms of its mission and landing locations (though it would not be able to perform orbital docking unless it was altered), and while initially designed for a single pilot, variants would have included a shuttle-style passenger area behind the cockpit that could have held four additional astronauts. It also had a payload bay capable of carrying 1,000 pounds.

And then, Robert McNamara, President Kennedy's secretary of defense, got involved—the same man who would later bedevil the MOL program. Formerly a president of the Ford Motor Corporation for a short time, McNamara brought what he considered to be a new style of doing business to the federal government. He was extremely analytical, and attempted to reduce war preparedness to quantifiable analysis. When it came to the air force's dream of a spaceplane, he had at first been conditionally supportive, renaming the program the X-20 and changing its mission from a suborbital one to an orbital program.[5] At the same time, he reclassified it as an experimental program. He seemed to change his mind often, bowing to political trends and numerical analyses. McNamara's odd attitudinal trajectory toward the X-20 can be seen in these relatively bland notations as referenced in the air force's own historical review:[6]

> February 23, 1962: Secretary of Defense, Robert S. McNamara, officially limited the objective of the Dyna-Soar program to the development of an orbital research system.
>
> July 13, 1962: Air Force headquarters informed ABDC headquarters that the Department of Defense had given qualified approval of the May 1962 system package program.
>
> January 18, 1963: The Secretary of Defense directed a review of the Dyna-Soar program.
>
> January 19, 1963: The Secretary of Defense directed a review of the Titan III program and the Gemini program of NASA.
>
> January 21, 1963: The Department of Defense and the National Aeronautics and Space Administration completed an agreement for defense department participation in the Gemini program.
>
> February 26, 1963: Headquarters of the Air Force Systems Command completed a position paper on the Dyna-Soar program, recommending continuation of the approved program.
>
> March 15, 1963: The Secretary of Defense directed the Air Force to conduct a comparison of the military potentials of Dyna-Soar and Gemini.
>
> May 10, 1963: Officials of the Space Systems Division and the Aeronautical Systems Division completed their joint response to Secretary McNamara's request for the military potentialities of Dyna-Soar and Gemini.
>
> July 3, 1963: AFSC headquarters informed the 1-20 office that the Department of Defense would only allow $125 million for fiscal year 1964.

A few months later, McNamara looked at two space-access systems (X-20 and Mercury/Gemini) and saw no real need for the former. While the X-20 was a very different design, aimed at accomplishing very different goals, he did not see the programs as sufficiently additive. He said that the X-20 program placed too much emphasis on high-tech and a steerable, gliding reentry when there was no real quantifiable reason for this capability. If the primary goal was to get men and weapons into space, NASA's ballistic capsule approach seemed sufficiently suited to the task. The same week he canceled the X-20 Dyna-Soar in December 1963 he announced the Manned Orbiting Laboratory program, which would give the air force access to space via NASA's technology, specifically the Gemini capsule.

When the X-20 was canceled, the final design had been settled on and was being prototyped. The spacecraft would have had a trans-stage rocket engine on the back for final boost and orbital changes, and capabilities, over time, as a suborbital bomber, high-speed test vehicle, reconnaissance platform, satellite deployment vehicle, and, possibly, armed Soviet satellite interceptor. In its ultimate form, the X-20 would have carried a total of five men, a payload and two robotic manipulator arms, allowing for both the potential disruption of enemy satellites and spacecraft, as well as the building and maintaining of a space station.

Systems that would have made their way into the X-20 were tested in other platforms. A few high-performance jet aircraft and the X-15 used similar maneuvering systems, and the X-15 pioneered the control of a winged craft from the airless, lift-less, near-orbital regime into denser atmosphere, transitioning from reaction controls (maneuvering thrusters) to aerial control surfaces (wing flaps or ailerons, rudders, etc.). The X-15 also tested high-speed landings on skids.

The X-20's ultimate purpose was never fully defined, but ideas included possible use as a nuclear bomber, replacing increasingly vulnerable air-breathing bombers; and as a replacement for the A-12/SR-71 high-altitude spy plane, which was expected to be vulnerable to enemy interception by 1969. The X-20 would be able to approach enemy targets with just a few minutes of warning as it dropped out of orbit or a hypersonic glide, much less lead time than the eighteen to twenty minutes provided by the tracking of an ICBM. An added benefit was that the spaceplane could also be called off if for some reason an attack was canceled. Dyna-Soar was expected to operate well into the 1970s.

When the program was canceled, the air force was, of course, unhappy. While some understood the apparent wisdom (and possible PR value) of merging their efforts with the Gemini program, the scientific and engineering returns from a program like Dyna-Soar were of a completely different nature and would be sorely missed. Even NASA opposed the spaceplane's cancellation, as the canny administrator, James Webb, suspected that the research and engineering not performed by the air force in the 1960s would have to be done by NASA in the 1970s and 1980s, and doubtless at greater expense. But McNamara did not heed the critics of the X-20's cancellation.

By the end of the Dyna-Soar program, $410 million in 1963 dollars had been spent.[7] The spacecraft itself never got past the mock-up stage, though it was ready to go into assembly the month the axe fell. But once McNamara had his final say, the program was disbanded, and the astronauts who had been in training were transferred to the X-15, the Manned Orbiting Laboratory, other air force duties, or NASA (including a young Neil Armstrong). X-20 research was carefully preserved, and was useful in the design of the space shuttle over a decade later.

Robotic boost-gliding spacecraft systems are being revisited by DARPA, with tests of various flight profiles and hardware iterations from the early 2000s on. None have worked to complete satisfaction, despite modern composite materials, sophisticated computer guidance and navigation, advanced flight modeling and the like—each flight test program has experienced significant challenges. So, given the technology of the 1950s and 1960s, it is unlikely that the X-20 and its ilk would have flown in the time frame and budget range the air force projected, if ever. One need look only as far as the space shuttle for another example of how difficult, and expensive, it can be to overcome the challenges of flying a winged craft into space and back. But had it been built and flown successfully, Dyna-Soar would have provided a whole different kind of human access to space, and a logical, ongoing infrastructure, well before the shuttle.

A wind-tunnel model of the *Silverbird*, Nazi Germany's suborbital "skip bomber" that would launch via rocket power into space, descend until it was able to "bounce" off the upper atmosphere, and glide to New York to drop large bombs on Manhattan. It never left the drawing board, but elements of the spaceplane found their way into US and Soviet space shuttles. *Image from NASA.*

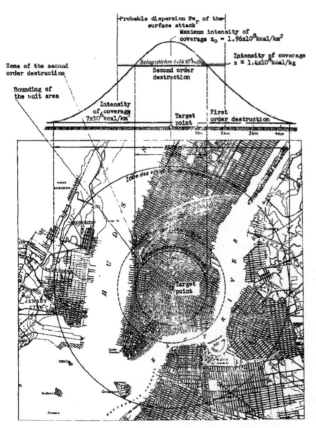

From Eugen Sanger's *Silbervogel* proposal, one of many maps of the projected bombing range of the rocket bomber. Manhattan was the target, and ground zero was within a few blocks of New York University. One of the munitions Sanger suggested for the skip bomber was a single 8,000-pound conventional bomb. *Image from US Department of Defense.*

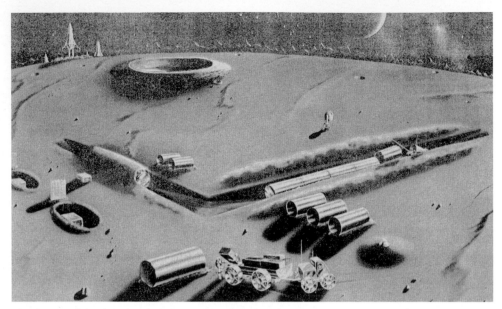

This is one of the few detailed drawings included in the Project Horizon study for a military lunar outpost. The study was completed in just a few months but was quite bold nonetheless. Shown are cylindrical habitation units to be buried under lunar soil for temperature control and protection against micrometeorites and radiation. *Image from US Army / NASA.*

Wernher von Braun used his first years in the United States to refine his thinking about spaceflight and exploration. One of the results was *Das Marsprojekt*, a slim but occasionally detailed outline of how the United States might set out to explore Mars. It was published in German first, around 1952, then later in English as *The Mars Project*. *Image from Wernher von Braun, Das Marsprojekt (Frankfurt: Umschau-Verlag, 1952). Cover illustration © Heinz Hähnel.*

After landing a single glider on the north pole of Mars, von Braun planned to have the crew assemble an overland tractor train to drive toward the equator. Once they arrived they would construct a landing strip to bring down additional gliders. Unfortunately, the Martian atmosphere was far too thin to support this winged glider design, but it would have been a sensational expedition. *Image from NASA.*

This contemporary NASA rendering of Orion shows a later variant, scaled down to work within the carrying capacity of the Saturn V booster. Regardless of the design variant, large or small, Orion's nuclear pulse drive would have offered far greater carrying capacity and reach than any chemical rocket. *Image from NASA.*

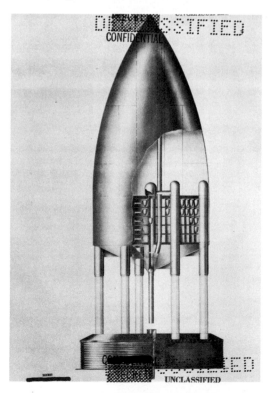

A cutaway illustration of the larger Orion variant. The bullet-shaped upper section is for crew and cargo, the midsection is the magazine for storing hundreds more small nuclear bombs, and below this, large tubular shock absorbers to moderate the sharp boosts from explosions impacting the pressure plate at the bottom. It looks crazy, but even contemporary engineers don't find major reasons that it couldn't work. It could have revolutionized the exploration of the solar system. *Image from NASA / US Department of Defense.*

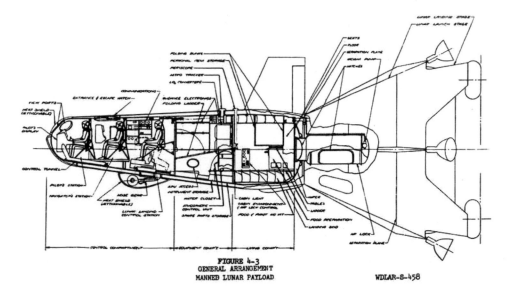

FIGURE 4-3
GENERAL ARRANGEMENT
MANNED LUNAR PAYLOAD

WDLAR-8-458

LUNEX was offered by the US Air Force as an alternative to Apollo. It was ultimately deemed impractical and duplicative of effort. *Image from NASA.*

The LUNEX launch facility was designed as a self-contained portal to the moon, complete with a capability for waterway deliveries of the spacecraft and booster elements. *Image from NASA/USAF.*

VON BRAUN'S
SPACE STATION
1952
Illustration by Chesley Bonestell

Von Braun's early 1950s design for an Earth-orbiting space station. This painting was created for NASA by the famed space artist Chesley Bonestell, who also illustrated much of the materials by von Braun and other scientists in the *Collier's* magazine series "Man into Space." *Image from NASA.*

MSFC-71-PD-4000-130

Shown here is the nuclear-powered variant of the EMPIRE Mars/Venus flyby ship. The power plant, at lower center, was a nuclear fission engine. The two arms at the top were to provide centrifugal force to create artificial gravity. The entire ship would rotate. *Image from NASA.*

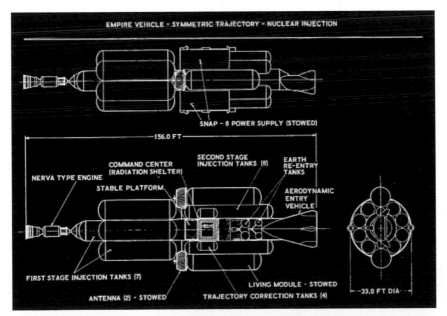

A schematic view of the design for Aeronutronic's nuclear-powered EMPIRE Mars/Venus flyby craft. While more difficult to engineer, the nuclear version was much less massive and more efficient. *Image from NASA.*

NASA astronaut Bob Crippen poses next to a Manned Orbiting Laboratory space suit, which he used during his time with that program. He later went on to fly four shuttle missions. *Image from NASA.*

The US Air Force's Astronaut Maneuvering Unit, or AMU. Designed to give US Air Force astronauts untethered mobility in space, the AMU was later transferred to NASA and flown on Gemini 9 for testing but was not used. A more developed variant of the AMU, the Manned Maneuvering Unit, or MMU, was tested in space in 1984. *Image from NASA/USAF.*

The Gemini capsule in orbit. Gemini was the first spacecraft that was fully maneuverable and able to rendezvous and dock with another spacecraft. This image is from the Gemini 6–7 mission. *Image from NASA.*

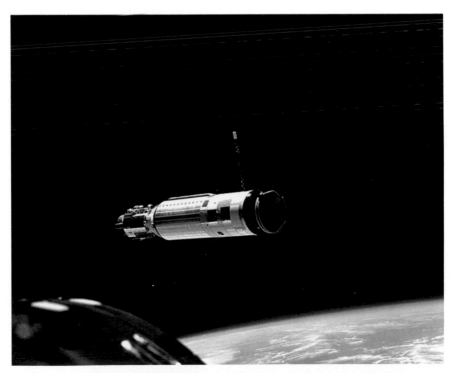

A view from Gemini 8's window during a fly-around inspection of the Agena prior to docking. *Image from NASA.*

Artist's concept from the 1960s of the Manned Orbiting Laboratory as it would have flown. The Gemini capsule has a hatch in its heatshield to allow the crew to transit from the capsule into the MOL without having to open hatches and conduct an EVA ("spacewalk"). *Image from NASA.*

Fabrication of the MOL mock-up at the McDonnell plant. MOL never flew with a crew aboard. *Image from NASA/USAF.*

Buzz Aldrin in the lunar module as Apollo 11 heads to the moon. Within hours he and Neil Armstrong would be standing on the lunar surface. *Image from NASA.*

Apollo 11's lunar module, taken shortly after landing. Inside the lower half, an ice plug that formed in a fuel line, causing excess pressure, had briefly threatened the mission. *Image from NASA.*

Artist's impression of the X-20 Dyna-Soar launching on a Titan II booster. The fins added to the base of the Titan booster were intended to stabilize the rocket during launch. *Image from NASA/USAF.*

This is a later version of the X-15, the A-2 variant, configured with external fuel tanks for longer flights. *Image from NASA/USAF.*

The X-15B would have been a stripped and strengthened version of the X-15 with a lengthened cockpit to accommodate a second astronaut. It was hoped that such a craft might reach orbit, but it was never built. *Image from NASA.*

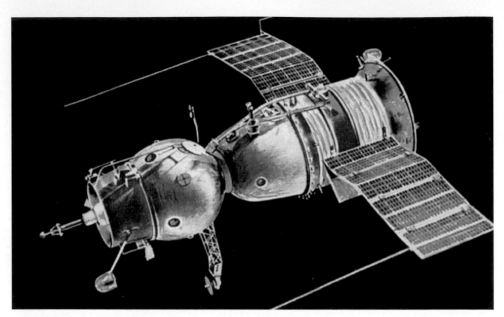

Soviet artist's impression of the early configuration of the Soyuz spacecraft. The spherical assembly to the left is the orbital module; the gumdrop-shaped module at center is the descent module. This illustration shows the solar panels fully deployed; during the Soyuz 1 mission, only one of the panels extended. *Image from NASA.*

Post-landing image of Soyuz 1. The circular structure at center is the hatch ring; little was left of the rest of the descent module after it burned. *Image from Roscosmos.*

The magnificent N-1, the Soviet Union's answer to the American Saturn V. This rocket, along with a Russian single-man lander, could have allowed the Soviets to reach the lunar surface sometime in the same time frame as Apollo, but the N-1 exploded every time it was flown. The program was ultimately scrapped a few years after it was clear the space race was over. Note the relative size of the men examining the booster at its base. *Image from TsSKB-Progress, http://en.samspace.ru/.*

Soviet-era image of the two turtles flown on Zond 5 around the moon. Other than losing a few ounces of weight (likely resulting from confusion about eating in space or lost water weight), they were none the worse for the experience. Just three months later, three American astronauts would repeat the experience, staying in lunar orbit for ten revolutions. *Image from RKK Energia.*

The SuitSat 1, a surplus Russian Orlan space suit, is released from the International Space Station with nobody inside. The test will monitor how long its electronics operate without cooling. A small radio transmitter was included to allow tracking by radio enthusiasts all over the world. *Image from NASA.*

SuitSat 1 prior to release from the International Space Station. The box affixed to the top of the helmet is the radio transmitter. *Image from NASA/Roscosmos.*

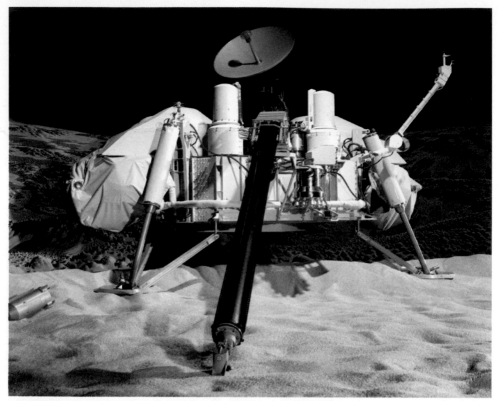

A mock-up of the Viking lander at a NASA field center, looking generally as it would on the Martian surface after landing. The small arm to the right is the meteorology boom, and the long arm to the center is the soil-sampling scoop. *Image from NASA.*

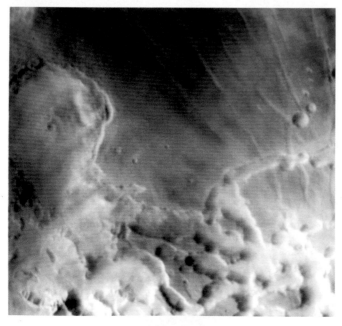

The Viking orbiters, with their upgraded cameras, provided vastly improved color images of the surface of Mars. This is a view of a region called Noctis Labyrinthus with morning fog. *Image from NASA/ JPL-Caltech.*

A panoramic view from Viking 2 of Utopia Planitia. Given that none of the rocks in the scene were visible from orbit, it's amazing that the two landers were able to land successfully under their own control. Today, JPL engineers refer to the Vikings with affection as "Big Dumb Landers" due to their primitive computers—but they worked. *Image from NASA/JPL-Caltech.*

Skylab as seen after emergency intervention by the first crew to arrive. The lack of a second solar panel is seen to the left. At the lower center, the corrugated-appearing unit is the cloth shade emplaced by the crew to shield the hull from the fierce effects of the sun. *Image from NASA.*

Alan Bean, commander of the second crew (of three) to inhabit Skylab, goes EVA outside the space station. Bean also flew to the lunar surface with Pete Conrad in 1960. *Image from NASA.*

Guy Gardner *(left)* and Mike Mullane *(right)* examine damage to the heat-resistant tiles on their shuttle orbiter after returning from STS-27 in 1988. This damage was caused by ablative materials falling from the top of the solid rocket boosters (SRBs), not foam from the external tank (which doomed *Columbia* in 2003). In either case, it demonstrates the variety of dangers the fragile tiles faced. *Image from NASA.*

Close-up view of the tile damage in the previous photo. The tiles in this area of the orbiter are very thick due to the temperatures they encountered, about 2,000°F. Charring can be seen in the area left exposed by the missing tile. *Image from NASA.*

"THE MEANDERINGS OF A WEAPON ORIENTED MIND
WHEN APPLIED IN A VACUUM
SUCH AS ON THE MOON" (U)

CLASSIFIED BY USA Weapons Command
JUNE 1965 TO GENERAL DECLASSIFICATION
SCHEDULE OF EXECUTIVE ORDER 11652
AUTOMATICALLY DOWNGRADED AT TWO YEAR
INTERVALS DECLASSIFIED ON DEC. 31 1971.

DOD DIR 5200.10

HEADQUARTERS
U. S. ARMY WEAPONS COMMAND
ROCK ISLAND, ILL.

The cover of the Rock Island Armory / US Army study of handheld weapons for use on the moon. Given the title and illustration, one imagines that it was intended primarily for internal use. *Image from US Army.*

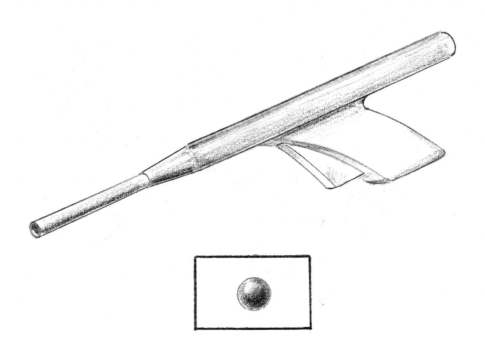

One of a number of designs proposed for a lunar combat pistol, the Spring Propelled Spherical Projectile pistol is also one of the "spacier" designs. It would fire a metal pellet intended primarily to puncture the space suit of a Soviet lunar soldier. *Image from Department of Defense / US Army.*

A model of the Soviet *Buran* space shuttle and its launch system. Unlike its American cousin, the winged orbiter did not carry large rocket engines. In *Buran*'s case, these were carried on the Energia booster system to which it was attached. *Image from Dreamstime.*

After its solo flight in 1988, the once-flown *Buran* orbiter was crushed when the hangar it was stored in collapsed in 2002. *Image from www.buran.ru.*

SpaceX, Elon Musk's private rocket company, has been providing resupply services to the International Space Station since 2012. They aim to fly astronauts in 2017 or 2018, and to land unmanned capsules on Mars soon thereafter. *Image from NASA.*

CHAPTER 13

BEYOND THE EDGE OF SPACE: THE X-15B

CLASSIFIED: *DECLASSIFIED IN 2011*

t was so close you could almost touch it. A steel-blue ribbon of atmosphere coated the edge of the Earth from up here, curving away toward the horizon. Some X-15 pilots claimed to be able to see north almost to Seattle, and south well into Mexico. And above all that ... blackness and stars, countless, unblinking stars.

Then he was over the top, the highest peak of the thrill ride now over, and the fifty-foot-long steel dart he was riding began to fall back toward Earth. It was time to take control again, fire the nose thrusters to get this metal stovepipe back into proper orientation for the long and hairy ride home, and land. It was a handful, but this was *flying*. And he loved it like nothing else.

Decades later, General Joe Engle, X-15 pilot, trained Apollo astronaut, and two-time commander of the space shuttle, summed up the first hypersonic rocket plane with simplicity: "The X-15 was the greatest airplane I've ever strapped my butt into, and I'll just say that right up front, it is the neatest machine to fly, the most professionally rewarding airplane to fly and climb out of, that I've ever been in."[1]

The X-15 program was the height of the X-plane experimental flying program in the United States. Beginning with the Bell X-1 in 1946 as a collaboration between the NACA (NASA's predecessor) and the US Army Air Forces (the US Air Force's predecessor), the X-planes pushed the boundaries of flight right to the bleeding edge. It was a low-budget, high-return program of daring, high-performance flying

169

that peaked with 199 flights of the X-15,[2] and would have continued to include orbital flights of the X-20 Dyna-Soar had the latter not been canceled in 1963.

Without the X-20 to cap the program, the US Air Force's highest and closest-to-orbit program was the X-15A-2 . . . and what an airplane that was. There was an even higher-performance variant planned in the late 1950s, the X-15B, which would have been a smaller and less capable version of the X-20, and was the brainchild of North American Aviation, the X-15's builder. If built, the X-15B would reach orbit, but first it would have to be approved and funded. As with so many space-age programs, especially those of the air force, it was not to be. But it could have been a glorious ride, if it could have been made to work.

The "standard" X-15 was fifty feet of mean.[3] The twenty-two-foot wingspan rocket plane was on the cutting edge of aerospace design when first flown in 1959,[4] and stayed at the pinnacle of piloted flight for the duration of its existence. Top-notch test pilots were carefully vetted for their ability to handle an aircraft operating right at the edge of the unknown. Into this select corps came Joe Engle, who began his exciting relationship with the X-15 in 1963.

The idea for a hypersonic test plane like the X-15 was first conceived by Walter Dornberger, a member of the group of German rocket men brought to the US by Wernher von Braun after WWII. Dornberger had been involved with German rocketry since the 1920s, and was appointed to develop rocket-powered weapons for the German military in 1930. He was well established in his post when von Braun was later recruited, and moved on Peenemunde and V-2 development in 1937. He was a powerful force throughout the war, and was a key proponent of keeping the program alive and well when it was threatened. Dornberger had headed a study for a hypersonic airplane for the NACA after arriving in the US, and this became the progenitor of the X-15. Dornberger was employed at Bell Aircraft in 1952, when the company began work on what would eventually become Dyna-Soar. The X-15 concept would be developed by others.

By 1954, a request for proposals (RFP in military/government parlance) had been sent to various aircraft fabricators, and North American Aviation was hired in 1955 to build the exotic craft. Another company called Reaction Motors was brought in to develop the rocket engine to power it in 1956.

The specifications called for a winged and piloted high-performance, high-altitude aircraft, powered by a rocket engine, to explore the realm of extreme speed (many multiples of the speed of sound) and the edge of space. The Germans had flirted with the idea of such a craft to be lofted by an advanced version of the V-2, but by the time development began in the 1950s, the X-15 was intended to be delivered to an altitude of about eight and a half miles by another aircraft—the B-52 bomber—at a speed of about 500 mph. During the short burn of its powerful rocket engine—there were two versions of this, one an eight-nozzle evolution of a small existing rocket engine, and later a large single engine—the X-15 would zoom to an altitude that the air force accepted as the beginning of space, fifty miles,[5] then glide, unpowered, back to a runway in the high desert of California. The larger of the two engines was fully throttleable, an innovation in rocket engines, and developed an ample thrust of 57,000 pounds.[6] It would land on ancient, dry, compacted-earth lake beds controlled by the air force, sliding to a stop on metal skids and a nosewheel.

The rocket plane was long and slender, with just the slightest suggestion of wings. It had a large, vertical stabilizer (a rudder to us civilians) that extended both above and below the fuselage—the craft looked like a pointed black pipe with fins. The bottom rudder extended lower than the rear landing gear did, and had to be discarded before the X-15 could land—it was equipped with a parachute and was recovered after each flight for reuse.

The body of the plane was constructed largely of an exotic alloy called Inconel-X, which was resistant to heat up to the limits of what the craft was expected to experience at maximum heating. Inconel-X was very tough to work with, being difficult to cut and weld, and even with its high heat resistance, it would begin to deform and melt if exposed to the maximum heating for too long. It was a case of "just enough" in terms of the protection it offered the plane and its pilot for the projected flight profile of the X-15.

Because the rocket plane flew above the majority of the atmosphere, the aerodynamic control surfaces on the wings and rudder would be ineffective for the highest part of the flight. To maintain control, maneuvering rockets, not unlike those used on spacecraft from the Mercury capsule through the space shuttle, were designed to allow the pilot to maneuver in the near vacuum of extreme altitude. The pilot interface, the control system, included a traditional control stick as was common in aircraft at the time, along with a smaller joystick to fire the maneuvering thrusters. It also sported a newly designed automatic control system to manage thruster firings and maintain attitude until the pilot reasserted manual control to bring the rocket plane back to Earth. A later evolution of this system controlled *both* the thrusters and the aerodynamic controls—ailerons and the rudder—and was operable for a much longer portion of the flight profile, when it worked.

The X-15 had a pressurized cabin, as did all high-altitude aircraft, but with a twist: the interior was pressurized not with breathable air (as in common airplanes), but inert nitrogen. This was much safer, as it would not support a flame inside the cockpit, but also meant that the pilot had to wear a pressure suit and helmet with a separate source of breathable oxygen. Once inside and ready to go, he could not lift his visor, lest he inhale unbreathable nitrogen.

The rocket plane also had an ejection seat that allowed the pilot to bail out up to speeds of Mach 4 (well below its maximum speed) and altitudes below 120,000 feet (well below its ceiling). It was an imperfect safety feature, but was the best that could be accomplished and covered most of the dangerous portions of the flight. The seat had its own set of retractable fins that would maintain its orientation in free fall until it was low enough for the pilot to deploy a parachute. Fortunately, this was never used at altitude.[7]

Most of the X-15's fifty-foot length was filled with fuel—anhydrous ammonia and hydrogen peroxide for the more powerful engine—and the rocket plane burned the 15,000 pounds of fuel in about a minute and a half. The scant remaining space was filled with control apparatus, monitoring gear, and, of course, one cramped human pilot. Two tiny windows allowed reasonably good visibility during flight, but because of poor visibility *below* the X-15, a chase plane was used during landing to guide the pilot to the dirt runway.

Each flight was different, and had specific goals. All began at the long runways at Edwards Air Force Base in the California desert. Joe Engle arrived early to watch the final preparations of his ride for that day. The low-slung rocket plane looked dangerous just sitting there, affixed to a pylon underneath the right wing of the B-52 that would haul it into the heavens. The entire X-15 program was lean and mean—it all looked very low rent compared to the activities occurring at the Cape, some 2,500 miles to the east. That was part of the beauty of it. It was just enough to get the job done, and done well, without too much fuss and bluster. In stark contrast to the huge launch complex, the multiple launch towers, the large storage buildings, water towers in Florida, and of course the control rooms taking form at the new Manned Spaceflight Center in Houston, this was all very simple. There were hangars along the runways at Edwards, and then the minimal clustering of the bomber, the rocket plane, a few trucks and jeeps, and the handful of technicians and officers preparing for the flight. That was about it. Oh, and the ambulance standing by not too far off . . . a not-so-subtle reminder that this was all a very risky thing for a young man in his early thirties to be doing.

Engle was already in his pressure suit, and once he completed a walk-around of the craft he was helped up the rolling stairway and into the cockpit. The canopy was lowered and sealed, and he was able to hear the oxygen flowing into his facemask as the cockpit began to pressurize with nitrogen. You did not so much as sit inside the X-15 as wear it; it was a snug fit for his six-foot frame. With the flight controller jabbering in his headset, he went through the checklist—all was well. The instruments were functioning fine and the controls responding to his input. It would not be long now.

The pilots in the carrier plane checked with him one more time, the jet engines just a few feet from his head began their shrill whine, the wheel stops were removed, and the bomber began its long, slow roll down the runway. Soon Engle's view changed from rolling runway to early-morning sky, and they were wheels-up.

As the flight program progressed, the project experimented with different configurations of the X-15. Some were specific research initiatives, but many were aimed specifically at increasing the performance of the rocket plane, expanding its speed and altitude. Over its decade of service, an experimental ramjet engine was fitted to the fuselage, huge tanks to accommodate extra fuel were strapped on, various coatings were applied to the hull, and improved flight control systems swapped out of the cockpit. Many records were set, broken, and set again. Top speed reached was 4,520 mph (Mach 6.7, almost seven times the speed of sound), and altitudes reached 354,200 feet, or sixty-seven miles.[8]

There were twelve pilots in the program. They were all test pilots but came from different backgrounds. There were names of fame and future fame among them.

Scott Crossfield was one of the first pilots to fly the rocket plane and the most experienced in experimental high-speed aircraft. After serving as a fighter pilot in WWII, he completed degrees in engineering and went to work for the NACA at Edwards. He flew the X-1 and its successor, the D-558-II, ninety-nine times.[9] He left the NACA to work for North American Aviation, the builder of the X-15, in 1955 and was the natural choice for early flights of the rocket plane—he was the company's premiere test pilot.

Neil Armstrong also served as a fighter pilot, flying for the US Navy in Korea. He too finished his degree after his military service, and also worked for the NACA. He left the X-15 project to join NASA's Astronaut Office in 1962, and ultimately reached the pinnacle of human spaceflight with the Apollo program.

And of course, there was Joe Engle. Engle flew with the air force after receiving a degree in engineering. He attended test pilot school on the personal recommendation of Chuck Yeager, the pilot who broke the sound barrier in the X-1. Engle flew the X-15 from 1963 through 1965, when he moved on to train for the Apollo program. While he never made it to the moon (he was replaced by geologist/astronaut Harrison Schmitt on Apollo 17), he eventually flew the space shuttle.

The X-15 was a sensory buffet even before it separated from the B-52. The small cockpit was flooded with the sounds of rushing air, the jet engine of the carrier plane, and the rattles and shakes of flight. Engle went through the checklist, twenty-seven critical steps, to make sure that the rocket plane was ready for flight. He activated the twin auxiliary power units, or APUs, that would provide function for critical processes. He then initiated the chill down of the rocket engine so that when the valves were opened the cold fuels would not "shock" the engine or cause the fuel flow to cavitate and be interrupted. He switched on the engine igniter, which determined his drop time—the spark plugs in the rocket engine could not run for more than about a minute without fuel or they would be damaged. The bomber crew signaled that everything was a go on their end, and Engle hit the drop switch. The X-15 fell free from the pylon, dropping quickly. The glide ratio of a recreational glider is about sixty-to-one (sixty feet forward for every foot of vertical fall); that of a 747 eighteen-to-one. For the X-15 it was four-to-1. It fell like a rock. He hit the engine switch, and the rocket ignited, the 57,000 pounds of thrust propelling the craft like an angry jackrabbit, and throttled the engine to 50 percent—it did not like any lower settings much. Pulling back the stick, Engle jammed the power to full and headed for the sky.

The X-15 program was successful from the start. There were issues. On one of Crossfield's early unpowered test flights, the rocket plane began a vertical oscillation and he barely landed it in one piece, slamming into the runway near the limits of the airframe. On the third flight, one of the eight small rocket engines exploded, and the rocket plane had to be landed with a full fuel load—something it was not designed to do. This resulted in hard landing that snapped the plane's backbone when it hit the runway, but it was repaired and flown again.

Later in the program, in 1967, X-15 number three went into a flat hypersonic spin upon reentry, carrying the pilot Michael Adams to his death.[10] The rocket plane broke up at around 60,000 feet, scattering wreckage across miles of desert hardpan. It was a dramatic reminder that this was a research craft, operating at its limits.

One of the three X-15s was modified to carry large auxiliary fuel tanks below the fuselage—they ran nearly the length of the rocket plane—greatly extending the burn time of the rocket engine. It was also coated with a new heat-buffering ablative coating, all in preparation for greater performance.

On today's flight, Engle was aiming for an altitude of 280,000 feet. The engine had to be switched off long before he reached the apex of the flight, since the momentum of the rocket plane would carry it far beyond cutoff. Running the large rocket engine just a few seconds longer than the flight plan intended could result in it being way off course—Neil Armstrong once glided all the way to the Rose Bowl in Pasadena, about fifty miles south of Edwards, before he got the craft turned around and directed back to the dry lake runway. You had to pay attention every second.

The X-15 was loud and aggressive. The thrust built up fast and just kept building. The fuselage would pop and bang as heat and aerodynamic pressures caused it to flex—the airframe would stretch, becoming almost two inches longer as the heat made the molecules in the alloy expand. The cockpit would sometimes fill with hot smoke. It was not a ride for the timid.

There was a simple stopwatch mounted to the instrument console. When it was time to shut off the engine the red sweep second hand pointed up and he killed the power—simple as that. Then the momentum carried it over the top past the apex of the flight, and he would begin the long glide back to a dead-stick landing.

For all the storm and fury of the X-15, it barely was able to scratch the underside of space due to its limited ability to stand the scorching heat of reentry. In theory, it could have gone much higher than it ever did, were it able to withstand returning to the atmosphere at the speeds required, up to Mach 25 (it normally came back to Earth at Mach 6). In contrast, even the early Mercury flights, with astronauts buttoned up in little capsules that were lobbed into the sky, like a cannonball, and returned to Earth—also like a cannonball—went higher and faster. But there was no real *flying* involved in the Mercury program—that was reserved for craft like the X-15. But in the endless competition for higher, farther, faster, the NASA guys were getting most of the glory. As usual, this did not sit well with some in the air force. Perhaps there was a way to wring more performance out of the rocket plane . . . maybe even to reach orbit?

One possibility was studied by North American Aviation in 1958, the X-15B. This would be a lightened X-15 that had been modified to make a one-orbit flight. It would not be dropped from a B-52 bomber, but would launch atop a set of rocket boosters, much as the Dyna-Soar had been intended to do. But since this proposal came from North American, the manufacturer, they proposed a booster of their own making to propel the X-15B into space rather than using another missile, such as the air force's Titan.

The choice was an interesting, if self-serving one. Immediately after WWII, North American had begun work on a launch system for a nuclear bomb based on von Braun's V-2 rocket. It was called the Navajo. But bombs were heavy, and the rockets small, so rather than try to build an Atlas or Titan missile, North American's proposal used a boost-glide design with wings, reminiscent of the Sanger *Silverbird*. The missile would launch vertically, then glide or fly under power like a high-altitude cruise missile. A number of variations were proposed, but all of these required a first-stage booster. The program continued for ten years, spawning engine designs that made their way into many other ICBM programs that moved forward to fruition. The Navajo did not, however. By the time it was nearing final tests, ICBMs were the kings of the nuclear weapons stockpile. A cruise missile like the Navajo took hours to deliver their deadly payloads; ICBMs took minutes. They were also harder to intercept than a cruising, winged missile.

But North American felt the extensive rocket engine development undertaken on the project should not go to waste. Having started with the V-2's engine, North American's engineers had worked on it for years, and in the end had a successful, large rocket engine called the G-38. This experience would help North American win multiple contracts for the Apollo program. But at the moment, with the Navajo canceled, there was a lot of unclaimed hardware left over. North American decided it might be able to put it to use—why not utilize these power plants to launch the X-15B? A cluster of three G-38s for the first stage and another for a second stage would provide just enough power to propel a lightened X-15 into a single orbit. This orbit would range from 250,000 to 400,000 feet (forty-five by seventy-five miles).[11]

With the rocket engine now silent and grumbles from the cooling power plant reverberating through the hull, Engle continued his ascent in near silence. The rocket plane continued to hurtle another 135,000 feet toward the highest point of its flight. He was now above most of the atmosphere, and there was little sense of motion or climbing, just a gradual variation of the view outside the windows as the craft slowly changed orientation. It was a stark contrast to the ninety seconds of powered boost. His flight path roughly bisected California, and he could see San Francisco off to one side, nestled on the curved edge of the Earth. He enjoyed the view for as long as he could, then fired the nose thrusters with the small hand controller to his left. The rocket plane was now falling back to Earth, and had to be in the proper orientation to reenter and glide back to home base. He would soon be very busy, but for now, the view—outer space was just ahead—was spectacular.

The gross liftoff weight of the combined launch system and X-15B would have been about 720,000 pounds, would cost $120 million, and take over two years to get flying by North American's estimates.[12] While capsule designs such as Mercury were simpler and possibly faster to accomplish, many regarded them as a dead-end path—other than carrying a human into space (and beating the Soviet Union to this goal), there was little else that they could do. The capsules would rocket into orbit and then plunge into the atmosphere, ending their flight by splashing down into the ocean, and that would be that. There was, at the time, little military application seen for ballistic capsules. A winged spaceplane, on the other hand, could be evolved into a number of roles, including providing a technological stepping-stone to the upcoming Dyna-Soar then under consideration by the air force. These types of vehicles seemed far more flexible, with the ability to perform military missions in space and land on runways, to be refurbished and reused. Importantly to the air force, there was also an ongoing and active role for a *pilot*—Mercury was largely automated to serve its simpler mission.

As the X-15 glided back into denser air, Engle transitioned to the right-hand controller. There was the traditional control stick in between his knees, but the pilots took pride in using the new hand unit to the right to control the aerodynamic surfaces. Soon the noises began again—the rocket plane was heating, and would soon reach its peak of about 1,200°F. This was accompanied with a symphony of bangs and groans as the hull heated and expanded, the joints of the structure sliding and grinding against each other as they were designed to do. The dry brown topography of central California could be seen below as he began the lazy curving trajectory intended to bring him back to Edwards. The flights were too short by half, he thought as he brought the craft down.[13]

The earliest mission profile designs for the X-15B were short and simple. The rocket plane would ascend on its stack of G-38 engines and then fire its own internal power plant to reach a low orbit. The successful flights had demonstrated its success at high speeds and in the upper atmosphere, right at the edge of space. There was still work to be done for the rocket plane to withstand the rigors of a full reentry, as the current X-15 only flew at about one-fourth the speed needed to attain true orbital ability. But these were simply problems to be worked out; at least that was what North American's proposal seemed to indicate.

In its basic form, the X-15B would hurtle into space, make a single orbit, and then glide into the upper atmosphere. Then the plan got bizarre—when the rocket plane had reached sufficiently low altitude, the pilot would eject and parachute to a landing, with the X-15B plunging into the Gulf of Mexico, not to be recovered. While this may seem wasteful, it was not much different than the one-use designs for Mercury and the Soviets' Vostok.[14] In all three cases, the machines were intended as single-use.

Design studies for the X-15B also considered the possibility of a two-place version, with a pilot and a "test director" riding in the backseat. This would have been a larger spacecraft, with an oxygen regeneration system and the ability to carry more payload. There would be a small payload bay to allow deployment of the experimental and reconnaissance hardware.

Any orbital plans for the X-15B would have required a profound evolution of many parts of the system, especially those to be reengineered for heat resistance. The regular X-15 was covered in Inconel-X alloy, which limited it to reentry temperatures of about 1,350°F, though some thought it could be modified to withstand nearly double that. The larger, higher-flying X-15B would be subjected to nearly 3,400°F on the nose and almost 5,000°F on leading-edge surfaces such as the front of the wings, and this would require *substantial* reengineering.[15] A potpourri of available but exotic materials had been considered for various parts of the skin, including columbium, thorium oxide beryllium oxide, molybdenum, graphite, tungsten, and René 41, the alloy used to construct the Mercury capsule. Merging these metals, with their different expansion rates under heating, would have been extremely difficult.

Engle came in via the usual path for an altitude flight. There were well-established glide routes to follow, one set for altitude flights and another for speed runs. The trick was to bring the X-15 down in the area you intended to, and he was not about to overshoot—the guys would never let him hear the end of it.

Not long before, in his first flight of the X-15, Engle had stunned his boss when he performed a barrel roll of the rocket plane during his final approach. He was not trying to be fancy; he was trying to scrub off excess speed and energy, and the X-15 stalled easily if a pilot got too pushy with the angle of attack. It made sense in the heat of the moment. A week elapsed before the maneuver was noticed when stunned superiors watched the onboard film. He got a brief grilling by his boss, Paul Bickle. When Engle explained that he needed to perform the maneuver to lose some speed, Bickle sympathized but asked him not to do it again.[16] If he did, all the other pilots would want to.

The most ambitious proposals for extended missions specified an X-15B capable of thirty-two orbits that could carry out an extensive research and reconnaissance test program. This might include optical and radio telescopic observations, and infrared, radar, and visual observations of the Earth, along with studies of human mobility in weightlessness, radiation and micrometeorite investigation, and biological research on the crew. Unlike the one-orbit flight profile, which would reenter as its energy was spent, the more ambitious flight plan would require some kind of retrofire capability to initiate reentry, using either the onboard engine or another system of solid rockets such as the Mercury capsule did.

Planned flight profiles would have pushed the orbital altitudes up to 600 miles, an extremely aggressive benchmark for a craft not originally designed for true reentry from orbital speeds of 17,500 mph. The final studies specified the use of von Braun's Saturn S-1 booster with 1.5 million pounds of thrust to lift the X-15B to orbit (a far cry from the middling power of the G-38 engine cluster) and a second stage to attain the higher orbits and lift the additional weight needed to maintain a crew in space for extended periods. These suggestions would, if implemented, have pushed the basic X-15 design well beyond what it was intended to do, or was likely to be suited for.

About as close as the program got to an orbital version was the X-15A-2, a high-speed experiment in add-ons and alterations. The rocket plane was lengthened by over two feet, large fuel tanks were strapped on to the fuselage, increasing the fuel load by 75 percent, and the hull was painted with an ablative, heat-resistant coating. It was flown to a maximum of Mach 6.7, but the special coating eroded due to high heat, and the A-2 was ultimately grounded. Clearly, a Mach 25 version of the rocket plane would require more effort than what made sense, especially since even with the needed improvements the basic design of the airframe might still not be up to the task.

After a couple of lazy figure eights, Engle was on the proper trajectory—no flyovers of the Rose Bowl for him. His instrumentation indicated that he was lined up for a proper landing. The X-15 fell quickly toward the long hardpan dry lake bed ahead.

A few feet above the ground, he toggled the landing gear and the skids and nosewheel dropped and locked. The rear skids hit first with a notable thud—at the speeds the X-15 landed, metal skis were needed at the rear instead of tires—and the craft pivoted quickly to bring the non-steerable nosewheel into contact with the runway. From there it was a two-mile slide until the rocket plane came to a halt, heat radiating from the fuselage. Within minutes the small fleet of recovery vehicles had arrived and Engle was able to climb out of the snug cockpit. Raising his helmet's visor, he savored the fresh air of the desert after breathing sterile, bottled oxygen. He then turned to admire the sleek form of his flying steed one more time—he never tired of looking at it—and headed off for a debriefing. Just one more day in the sky.

Ultimately, engineering challenges and the changing nature of the space race killed the X-15B's chances of reaching fruition. If built, it would probably not have flown into orbit before 1964 or (very likely) later. This would have been a date well after John Glenn's 1962 orbital flight in NASA's Mercury program, which would have accomplished the immediate goal of "catching up with the Russians." While an orbital flight of the X-15B (or the X-20 Dyna-Soar for that matter) would have been an entirely different level of achievement from the Mercury program, it was not deemed necessary by the federal government.

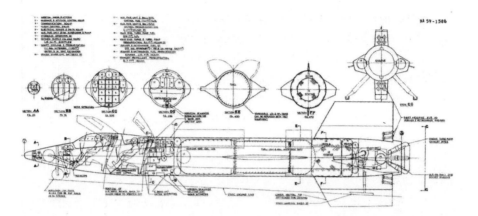

Fig. 13.1. A cutaway view of the X-15B orbital spaceplane. Note the second crew member, seated snugly behind the first, to left. The X-15B would have been, in many ways, a new aircraft built on designs proven for high-speed flight, but not orbital reentry. Image from NASA.

Even without an orbital variant, the X-15 program excelled and flew until 1968. The rocket plane developed and proved many technologies critical to hypersonic flight and paved the way for the space shuttle in the 1980s. Advanced versions of the X-15, equipped with the external fuel tanks intended to fly on the X-15B, increased its performance greatly, pushing the envelope of what the rocket plane was capable of. It was a wildly successful effort on a shoestring budget.

And what did the pilots think? Joe Engle cited the X-15 as one of his favorite flight programs (impressive, coming from a man who flew the space shuttle), and Scott Crossfield famously said that the X-15 was one of the few aircraft that could cause grown men to cry upon its retirement. That says it all.

CHAPTER 14

THE SAD, STRANGE TALE OF SOYUZ 1

Vladimir had known things could end this way. For months, the program had been in shambles, with abrupt changes in staff, pressure from the Communist Party, and a spacecraft plagued with problems—Yuri Gagarin, the first man in space and his close friend, catalogued over 200 critical faults during a single inspection. Gagarin, a bemedaled Hero of the Soviet Union, had offered to take Vladimir's place, presumably to then refuse to pilot this flying death trap and protect them both from retribution. But in the end, party loyalty and a sense of mission had prevailed, and Vladimir agreed to make the first Soyuz flight, despite a deep sense of foreboding. Now he was hurtling through space, barely controlling the failing, defective spacecraft, and headed for reentry whether he was ready or not. . . .

Damn the Soviet Union, damn the Communist Party, and damn the Soviet leader Leonid Brezhnev, who had forced him into the horrible machine before it was ready. He yelled into the radio, cursing the party, short-sighted politicians, and this "devil ship" that was carrying him to his doom. His last transmissions expressed his regrets to his beloved wife, until the ionization of his fatal reentry made communication impossible. . . .

That's one version of the sad story of Soyuz 1, the Soviet Union's challenger to beat Apollo to the moon. Other, less dramatic (and more reliable) accounts have the lone cosmonaut, Vladimir Komarov, stoically working on the problems and calling out data in a calm voice until the Soyuz's signal faded.

When dealing with space race history from the perspective of the

Soviet Union, it can be tricky to find a version that the majority of historians can agree upon. What *is* known for certain is that the first flight of the Soyuz spacecraft took place before the putative moonship was ready, that the short mission was plagued with problems, and that it ended in disaster—the parachute malfunctioned, and the spacecraft plunged into the Russian wilderness at almost 100 mph, leaving flattened, flaming wreckage, and little else, in its wake.

Upon deeper examination, the story of Soyuz 1 seems emblematic of the entire Soviet moon effort, especially after the death of its chief designer, the Russian answer to Wernher von Braun, Sergei Korolev. Korolev was the enigmatic genius who had led the USSR's space program from its triumphant origins through his premature death, which was accelerated by exhaustion and a punishing schedule dictated by the Soviet leadership, in January 1966. Despite early run-ins with the Stalin regime that had landed him in a labor camp for six years starting in 1938,[1] Korolev's genius was widely recognized and he later became a shining star in the Russian space establishment. When it was time to repurpose an R-7 ICBM to launch Russia's first satellite, Sputnik, the leadership turned to Korolev. The product of his labor, Vostok 1, carried Yuri Gagarin into space for a single orbit—a resounding first in the space race.[2] Korolev then engineered the first spacewalk by Alexi Leonov (a successful but near-fatal excursion) in 1965. All these accomplishments came before their American equivalents, leaving the US in a position of puzzled embarrassment, wondering aloud how the Soviet Union, which had been in postwar shambles just over a decade earlier, could have come so far, so fast. The answer was a complex one, but hinged on successes in brute-force rocketry and the ability to pick the low-hanging fruit of spaceflight—the near-Earth firsts—quickly and effectively. When President Kennedy announced the US goal of sending a man to the lunar surface in 1961, it was intended to be a technological equalizer, and the USSR's space program had to ramp up quickly to refine its technologies and attempt to meet the US challenge. In one sense, the Soviets almost won the race to the moon—had they been more willing to gamble with human lives than they were, they could probably have sent cosmonauts in a single loop past the moon before Apollo 8 orbited that body in December 1968. Such a mission, had it successfully brought the crew members home alive after their lunar flyby, would have stolen much of America's thunder. But it was not to be, and the fiery end to Soyuz 1 presaged many other failures of their lunar effort.

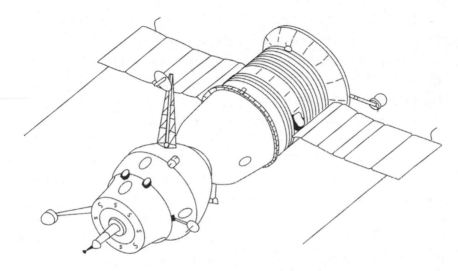

Fig. 14.1. Schematic view of the early Soyuz configuration. In Komarov's flight, one of the solar panels failed to deploy. With a number of revisions, the basic Soyuz design has flown into space many hundreds of times over sixty years. Image from NASA.

When the USSR embarked on its quest to put a man in space after initiating the space age with Sputnik, the Soviets knew they needed to move fast. They already had a powerful rocket in the R-7, their ICBM in use since 1959. Now they needed a capsule to orbit a man in space. That project began under Korolev just a few months after the first Sputnik launches, with the limitation that the existing model of the R-7 could only carry five tons.[3] They also wanted an all-purpose spacecraft that could house robotic spy equipment for orbital reconnaissance of the US. Their answer was the Vostok space capsule, a 10,000-pound (roughly two-and-a-half-ton) machine comprised of a spherical pressurized hull about seven feet in diameter, with a lower stage containing maneuvering fuel, retrorockets, and other equipment. Studies had shown that a capsule with a blunt bottom, such as the American spacecraft were employing, would be more efficient and less stressful during reentry—the blunt-body design allowed it to glide a bit, instead of simply plunging into the atmosphere—but time was short and the Soviets wanted to beat the US above all else. A spherical design would work, and that was what counted. The Vostok became the workhorse of their space efforts from 1960 through 1963.

Vostok was a masterpiece of simplicity. The center of mass was offset so that however it was oriented when it reentered the Earth's atmosphere, the capsule would default to a shield-down orientation. Electronics and guidance systems were simple and basic, with redundancy built in. Most systems were automated, but had manual backups (as did their American counterparts), and the knowable contingencies were accounted for. An example of this ingenious and simplistic design was the manual reentry device, a multi-lensed periscope that extended from the hull. When the spacecraft was on the daytime, or sunlit, side of the Earth, and in the correct orientation for reentry, all the lenses would be flooded with sunlight, indicating that the spacecraft was properly aligned to fire its retrorockets. Alignment along its orbital path was assured by matching up marks on the viewing glass with the passing landscape below. It was foolproof and sturdy. An automatic reentry device could also be used, utilizing basic star-tracking photocells that were tied into an analog computer; this would work in sunlight or on the night side of the Earth equally well.

For landing, the Vostok used a parachute just like the US Mercury capsules did, but unlike Mercury it was not intended to make a *soft* landing. Instead, Vostok capsules had an explosive hatch and an ejection seat. Rather than ride the capsule into a survivable landing in the sea as the Americans would, the cosmonaut would blow the hatch at an altitude of a mile or more and eject using the modified jet fighter ejection seat. The capsule would land empty with a sharp crunch on the steppes below. Again, it was basic and functional.

There was one major variant of the Vostok, the Voskhod capsule that flew in 1964 and 1965. By then, the US had graduated from the simple one-man Mercury capsule to the two-man Gemini spacecraft, which had true maneuvering capability to afford rendezvous and docking exercises in orbit, critical to the Apollo lunar landing efforts. But with two astronauts flying in Gemini, the USSR had to up the ante, so the Vostok design would be modified to carry not two but three cosmonauts. It was a dangerous choice. The basic design remained, but the ejection seat was removed and three slim couches installed— the fit was so tight that the cosmonauts would be unable to even wear pressure suits. Braking rockets were added at the base of the space- craft, and would be triggered at the last moments of descent to prevent the occupants from being injured during landing. Two missions were flown with this variant—first the three-man mission, and then, with one seat removed to accommodate an inflatable airlock, a two-man crew performed the first spacewalk of the space race. It was a spec- tacular achievement, albeit a dangerous one (the cosmonaut's space suit puffed up so much that he was unable to reenter the airlock until he vented air pressure—a dangerous procedure in space). But with no rendezvous or docking capability, Voskhod was a dead-end design, and the Soviets knew it. In 1962, with Kennedy's lunar taunt echoing through the Soviet bureaucracy even as the Vostok was being modi- fied into Voskhod, work began on an entirely new design that would become the Soyuz.

Vladimir Komarov joined the USSR's first group of cosmonauts in 1959. He had trained as a combat pilot during WWII and barely missed being sent into action when the conflict ended in 1945. By 1960 he was an experienced pilot and engineer, and had participated in the design of Soviet spacecraft. In October 1964, he was one of the brave trio to board the Voskhod spacecraft, sans pressure suit, to fly the only three-man mission until Apollo 7 in 1968.

Komarov was then assigned to the Soyuz program. Like their American counterparts, the cosmonauts were involved in overseeing the design and manufacture of their spacecraft, and Komarov, with his engineer's eye, was not happy with what he saw. He lodged formal complaints about design and operational issues with the first Soyuz capsules, including the fact that he felt that the spacecraft's hatch was too small for a fully suited cosmonaut to exit or enter while weight-less.[4] There was little response from his superiors, and Komarov reportedly went so far as to have Gagarin, the number one hero in the Soviet Union, bring up the issue with Leonid Brezhnev. But the hurried, shoddy work on Soyuz continued.

In 1967, Komarov was selected to pilot Soyuz 1, with Gagarin as his backup. The two labored for long days, often working through weekends, to prepare for a mission they both knew was dangerous. But that mission was critical to keeping the Soviet Union ahead in the space race, and nobody was going to step aside and allow a comrade to fly in his place. The training continued, even as the issues mounted.

Then, in late January 1967, the US program had a terrible accident that should have given the Soviets a breather. The Apollo 1 capsule, undergoing final preflight tests with three astronauts aboard, caught fire on the pad, killing the crewmen instantly.[5] The Apollo program came to a sudden halt while the cause of the fire was investigated. Despite the political differences between the two countries, Komarov and his fellow cosmonauts felt that brothers across the Atlantic had been lost. Due to the transparent nature of NASA's endeavors, it was clear that NASA would stand down on further flights until the problems were solved—in the end, almost a year and a half. This should have given the Soviets time to slow down the Soyuz program, reexamine their hard-ware, and schedule increased testing and development, but this is not what occurred. The program continued at the same breakneck pace as it had, and problems with the spacecraft were simply shoved aside. It would fly in 1967. One of the Soviet Union's biggest holidays was coming up, the Day of Worker's Solidarity on May 1, and the flight of Soyuz 1 would be a perfect public relations coup for the "worker's para-dise." The Soviets had also not flown a man into space since Voskhod 2, two years earlier, which was embarrassing given the flight rate of the American Gemini program. Soyuz 1 would fly on schedule.

Plans drafted in 1962, then redrawn in 1964, portrayed the Soyuz as a multipurpose craft—the USSR would not be following the sequential steps of NASA's Mercury, Gemini, and Apollo programs. Nor, as it turned out, would it abandon the space hardware it developed at such expense after 1975 as the US did; Soyuz continues to fly in the twenty-first century. Soyuz was envisioned as a Gemini competitor, a spacecraft that could change orbits and perform the rendezvous and docking procedures required for a lunar landing effort. It would then become a lunar spacecraft, flying around the moon with crews aboard. Finally, Soyuz would be paired with a Soviet LK lander, a small single-man version of Apollo's lunar module, to reach the moon and plant a Soviet flag. It was intended as a modular system that would evolve as the requirements did—including military needs. It is still evolving today, even as a replacement is being planned.

There was unrest and competition in the USSR's lunar program, however. Vladimir Chelomei, another Russian rocket genius and ICBM designer, ran a competing design bureau and had his own plans for lunar flight. When the first Soyuz designs were being put forth in 1962, Chelomei's moonship plans were adopted instead. It was a smaller one-seat spacecraft intended for a lunar flyby, leaving Korolev's Soyuz to evolve in the background as the eventual landing program.

But this all changed with the removal from power of Premier Khrushchev in 1964—coincidentally, right in the middle of a Soviet space mission—and with the shift in power, Korolev found himself back in the first position in the Russian space effort with the Soyuz in the lead. While the Soyuz was still intended, publicly at any rate, as an orbital rendezvous and docking system, Korolev had his engineer's eye squarely on the moon. In October 1965, Korolev was in control of the circumlunar flight project, but died of cancer a few months later. He left behind an entire space program—the Soyuz and a lunar-capable superbooster called the N-1, which was generally equivalent to von Braun's Saturn V. And much of the program was a mess.

Unlike the US program, the Soviet lunar plan required multiple launches and crew transfers from one spacecraft to another via

EVA—the cosmonauts would actually have to exit one spacecraft and maneuver themselves through open space to another. And even this somewhat ad hoc design required the large N-1 booster to succeed, a lunar lander to be tested, and of course the Soyuz had to prove itself capable for the task ahead. Work continued after Korolev's death under his lieutenant in the design bureau, Vasily Mishin. Mishin was a competent engineer, but did not have the political gravitas of Korolev, and the lunar program began to suffer further under his control. With Chelomei continuing to advocate his own systems, and with constant political pressure from the Communist Party as well as many technical problems—especially with the giant N-1 rocket—Mishin was unable to keep the program moving at a pace needed to beat the Americans. Perhaps nobody could have—Korolev died before it was really possible to see all the challenges facing the program. The N-1 was perhaps the most vexing of all; it failed in every launch attempt and was eventually canceled outright, dooming the Soviet lunar effort even as the Apollo landings proceeded to explore the moon.

But little of this was known in 1967. What Vladimir Komarov did know in the months leading up to his flight of Soyuz 1 was that the Soyuz program was plagued with problems, and these issues were more than just mechanical and electrical faults. The leadership of the space agency, and higher, seemed deaf to the concerns of the cosmonauts who would be flying these machines into glory or disaster. And as fatalistic as one tended to be in the USSR's military forces, this was a step too far. But there was nothing to be done but proceed.

In November 1966, the first unmanned orbital launch of a Soyuz—called Kosmos 133 for testing purposes—was a failure, with dozens of mechanical and electronic faults indicated during launch and ascent. It spent two days spinning in orbit, as attempts to regain ground control continued to no avail. It was impossible to ascertain where the spacecraft might impact after the eventual uncontrolled reentry, but Chinese territory seemed likely. The spacecraft was finally destroyed before it could reenter, leaving a cloud of fragments and detritus in its wake.

In December another unmanned orbital launch was attempted

with an identical Soyuz. This spacecraft had originally been intended to launch a day after the Kosmos 133 flight and perform an automated rendezvous, but that launch was scrubbed when the first spacecraft spun out of control. A second launch attempt was shut down immediately after ignition by a condition called underthrust, in which the engines do not produce enough power. The indication turned out to be faulty, but the onboard Soyuz escape system, which, like American spacecraft, used a smaller set of rockets above the capsule to pull it away from the booster in an emergency, thought that the booster was malfunctioning in flight and activated while still sitting on the pad. The fully fueled rocket was a bomb waiting to go off as the Soyuz capsule rocketed away, and the booster exploded, ruining the launch facility and killing a number of engineers and technicians.[6]

But rather than stopping the program to investigate these failures, Soviet leaders insisted that the first manned flight proceed on schedule. It was unthinkable, foolhardy, careless, and criminal in the minds of the cosmonauts, but there it was—orders from the top would force them into success or flaming oblivion. Komarov's Soyuz 1 mission was scheduled to fly in just a few months.

What makes the aggressive schedule even more ironic in hindsight are the original intentions for Komarov's flight. Rather than a simple test flight of the Soyuz, as had been described by Soviet media after the tragic end of the mission, the original flight plan had been remarkably similar to that of the unmanned failures that preceded it, and some of the Gemini missions of 1965 and 1966. Soyuz 1 was to launch with Komarov at the controls, and once he was in orbit with all systems checked out, a second Soyuz would launch with three cosmonauts aboard. The two spacecraft would rendezvous and dock.

But no matter the mission specifics, the overall idea of two Soyuz spacecraft launching, rendezvousing, and docking seems absurd given that an almost identical unmanned mission profile had just failed—not with one or two small issues, but with two spectacular explosions. Yet the Soviet leadership now demanded that cosmonauts take these dangerously flawed spacecraft into orbit and attempt the same thing, but this time with lives at stake. It was insanity, fueled by the space race and recent rendezvous and docking achievements by the Gemini program.

As one seeks more details of the decision-making process behind rushing the ill-fated flight, dependable sources become thinner. But more than one reputable account cites direct pressure from a member of the Soviet politburo, Dmitry Ustinov, on Mishin, Korolev's replacement. There was a political summit coming up, with leaders of the Soviet bloc countries slated to meet in Czechoslovakia, and Ustinov wanted the flight to occur on or near the eve of the summit. It is also alleged that Ustinov spoke with Komarov directly, threatening a reduction in rank or worse if he did not fly. Ultimately, the department heads under Mishin met after the most recent test flight and, with some reluctance, agreed to proceed.

As planned, Soyuz 1 would launch on April 23 and establish stable operations in orbit. Komarov's spacecraft would carry an active docking mechanism. Twenty-four hours later a second Soyuz, this one carrying three cosmonauts, would head into orbit carrying a passive docking mechanism. Once both craft were fully operational, they would rendezvous, and Komarov would complete the final approach and docking. By some accounts, two of the three cosmonauts in Soyuz 2 would then exit the spacecraft, make an EVA to Komarov's Soyuz, and return to Earth with him. The final lone cosmonaut would bring Soyuz 2 back to Earth, completing the heroic mission ... and giving Soviet leaders bragging rights at their conference.

As late as April 14, project management was still reporting many issues with the spacecraft, but these were ignored. Mishin, doubtless under tremendous pressure, reportedly flew into a rage and threw the messenger out of the meeting.

At this point, Komarov and Gagarin were in their final training runs. Their concerns were mounting, but still there was no response from their superiors. Here the versions of events diverge again: some sources claim that just before the flight, Gagarin insisted that he be given a space suit and that he would fly the dangerous ship—his alleged logic being that as a supremely important VIP in the Soviet PR machine, essentially a national treasure, the leadership would scrub the mission rather than let him fly. Other versions relate that Gagarin offered to step up (possibly expressing the same logic to Komarov), but that Komarov declined the offer, concerned that Gagarin would be forced to fly the mission and be killed. Both accounts are heroic and in keeping with the myth-building style of the USSR, but admitting to either version would have severely tainted the upper levels of government and military, and neither account has been reliably confirmed.

Soyuz 1 launched with Komarov aboard on April 23, 1967. The ascent proceeded as expected, doubtless engendering many sighs of relief from flight controllers and Komarov himself. Less than ten minutes after launch, reports indicated that the Soyuz had separated from the booster and was in orbit—welcome news. Then the mission began to unravel.

The Soyuz spacecraft, unlike their American counterparts, were dependent on solar panels for power. US spacecraft from Gemini through Apollo were powered by chemical fuel cells, which required only the opening of a few valves to function. Soyuz had two large solar panels that deployed from each side of the spacecraft, a configuration they retain to this day. Within minutes, telemetry from Soyuz 1 indicated that the left solar panel had failed to unfold, and that electrical power was far lower than needed for the planned mission. In his second orbit, Komarov confirmed the failure and was said to have resorted to kicking the inside of the capsule in a vain attempt to free the stuck panel.

Additional failures were noted on a lengthening list. The backup telemetry antenna had not deployed, restricting the flow of data needed to monitor the spacecraft's systems and status. Then, ground controllers noted that the cover on a critical star and sun sensor had not opened due to being blocked by the undeployed solar panel. This would prevent Komarov from having the positional data he needed to navigate or initiate the desired spin of the craft for stabilization.

Komarov soon confirmed this, and added that the maneuvering fuel readings were low and the spacecraft was not responding properly to attitude control firings—the mass was off-center due to one long solar panel sticking out of one side but not the other.[7]

Ground controllers wanted to terminate the mission while there was enough power in the batteries to do so—this would soon be depleted by the electronics in the spacecraft. But they were overruled by party leaders, who wanted Komarov to continue in his attempts to manually reorient the Soyuz and continue the flight.

Komarov attempted to comply. He hoped to recharge the rapidly depleting batteries, but the single solar panel was only delivering about half the power needed to do so, and the attempt was fruitless. He continued to burn his limited supply of attitude control fuel to orient the spacecraft properly, but this too proved impossible. A few officials continued to press for further attempts to prolong the mission, but the engineers kept the pressure up—there was no way to recover from this many failures. Only within a half hour of the point of no return was final permission to terminate the flight given. On Komarov's sixteenth orbit, Gagarin was on the microphone in ground control, radioing up instructions and data for the reentry attempt.

As Soyuz 1 entered its critical seventeenth orbit, Komarov radioed what the tracking data had indicated on the ground—the spacecraft was not properly set up for reentry. It was drifting from the course he had finally managed to set, and the controls were still not working correctly. The faulty spacecraft systems were now in a cascading failure.

Controllers prepared a new reentry plan for the eighteenth and nineteenth orbits, knowing that Komarov would be on the last dregs of battery power by then. They instructed him to attempt to align the spacecraft while on the daylight side of Earth, then set the onboard gyroscopes and hope that they would maintain proper orientation long enough for retrofire. The new instructions were not something that Komarov had trained for, and it was going to be touch and go, but he agreed to try.

Communications were sporadic, but Komarov did manage to convey that the braking engine had fired, only to be shut down early by an errant computer command. As the spacecraft drifted out of the proper orientation during the retrofire, the navigation system had sensed the deviation and stopped the engine. But according to their calculations, the firing had lasted long enough to cause the spacecraft to reenter. Soviet ground forces were redirected to the new projected landing point. Defense radar soon detected the spacecraft in the upper atmosphere, but could tell little else. One way or another, it was coming down.

At this point, information ceased to come into the control center, and repeated queries to the recovery team were greeted with silence—controllers could only wait and hope. On-scene, one helicopter reported sighting the descent module in the air, then on the ground. It was tipped on its side with the parachute billowing across the open plain. As the helicopter descended, crew members saw the landing rockets fire—but this should have occurred before the capsule reached Earth. Whatever slowing effect they would have had, which was critical to Komarov's survival, was wasted as they blasted uselessly into the air adjacent to the crashed capsule.

The helicopter landed a few hundred feet from the column of dark smoke that now marked Soyuz 1. As the technicians and medical personnel rushed to the landing site, it became clear that there was no hurry—there could not possibly be anyone alive inside. The interior of the capsule was burning, the flames visible via a large hole that was burned through the hull. Molten metal dripped onto the hard soil.

In a tragic coda, rescue forces still in the air saw a flare fired into the sky by a recovery worker—the color code indicated that the cosmonaut needed medical assistance. There was no code for "cosmonaut deceased," and the resulting reports of a medical emergency generated more confusion.

Fire extinguishers were insufficient to battle the hot flames, and rescuers shoveled dirt onto the capsule. With one shovelful too many, the brittle, torched spacecraft collapsed flat, leaving only the stronger hatch structure standing. Would-be rescuers then shoveled dirt off the wreckage, only to find the charred, unrecognizable remains of Komarov in the pilot's seat. He died well before he burned, due to an impact speed of nearly 100 mph.[8] The news finally reached the stunned teams back at ground control, and the room went silent. The mission of Soyuz 1 was officially, and tragically, over.

A postmortem commission investigated the scant remains of the spacecraft and ascertained that the parachutes had failed. The main chute had snagged in its enclosure, and a backup chute, deployed automatically, also snarled as the capsule made its twisting, horrific descent. Gradually more evidence emerged, though it remains controversial: there were indications that glue-like materials in the protective coating on the capsule had made their way into the parachute container, causing the parachute to stick to the walls of the enclosure. A spacecraft-baking process used as a final step in preparing the thermal protection barrier had caused the fatal seepage. In the end these clues did result in a safer spacecraft design and assembly process, but they came too late for Vladimir Komarov.

The dramatic failure of Soyuz 1 caused a shutdown of the USSR's manned spaceflight program for almost the same duration as that experienced by the US in the wake of the Apollo 1 fire—both programs were suspended for about eighteen months as processes and procedures were reviewed and improved. These staggering losses—for the US, three astronauts killed on the launchpad by an oxygen-fueled fire, and for the Soviets a cosmonaut killed when his spacecraft experienced an almost unthinkable series of malfunctions that dogged it all the way to the ground—were cause to question all assumptions about how the two superpowers went about spaceflight. The basic designs of both spacecraft systems were sound, but the headlong rush to outpace their competitor caused reckless behavior on both sides. The resulting improvements doubtless saved lives in both space programs.

Yuri Gagarin mourned his fallen comrade as did the other cosmonauts and associated personnel. Komarov's death was observed in a manner accorded to the Soviet Union's greatest heroes, and his remains—by numerous accounts, only a small, charred clump—were interred in the Kremlin Wall Necropolis in Moscow. He was posthumously awarded the Order of Lenin and made a Hero of the Soviet Union, like his dear friend Gagarin (who himself died in a jet crash in 1968[9]). On July 20, 1969, Neil Armstrong and Buzz Aldrin honored Komarov, Gagarin, and the Apollo 1 astronauts in what was possibly the most fitting way, with a small memorial package left on the lunar surface during their moon walk.

Fig. 14.2. Vladimir Komarov, Hero of the Soviet Union and first cosmonaut to fly the Soyuz spacecraft. The reckless program cost him his life. Image from NASA.

The day after the crash, an officially approved open letter by Komarov's fellow cosmonauts was published in the USSR's primary news outlet, the newspaper *Pravda*. It said, "For the forerunners it is always more difficult. They tread the unknown paths and these paths are not straight, they have sharp turns, surprises and dangers. But anyone who takes the pathway into orbit never wants to leave it. And no matter what difficulties or obstacles there are, they are never strong enough to deflect such a man from his chosen path. While his heart beats in his chest, a cosmonaut will always continue to challenge the universe. Vladimir Komarov was one of the first on this treacherous path."[10] Gagarin later commented, "He has shown us how dangerous the pathway to space is. His flight and his death will teach us courage."[11]

The Soyuz spacecraft, much improved after its first ill-fated flight, was redesigned and in various incarnations has been in continuous service since that time. In 1975 it performed a rendezvous and docking with the last Apollo spacecraft flown, a remarkable demonstration of détente between the US and USSR. Soyuz flew countless missions to build and supply Soviet space stations, and has continued to be a workhorse for the International Space Station in both manned and robotic forms. With its purpose-designed Soyuz booster, it has become the most reliable and frequently flown crewed spacecraft in history, an achievement that would surely have gladdened Komarov had he survived his perilous mission.

CHAPTER 15

THE TURTLENAUTS

CLASSIFIED: *DECLASSIFIED AFTER THE FALL OF THE SOVIET UNION IN 1991*

From 1961 until 1969, the USSR and United States were in a race to land a man on the moon. From the time President Kennedy announced his intention to chart a course to the lunar surface, the CIA and NSA devoted increasing amounts of energy trying to ascertain Soviet intentions and progress. The reports throughout the 1960s were uncertain but sometimes prescient—the intelligence-gathering operations were closer to the truth than could be known at the time. The Soviets continued to maintain that they were not engaged in a footrace to beat the US to the moon, but as records of the USSR's space efforts have become available since the fall of the Soviet Union, it is clear that they were undertaking substantial, though sometimes conflicting and inconsistent, efforts to reach the moon.

Critical to this effort was increasing the size and capability of their rockets. Despite their early lead in rocket capability and orbital achievement, by 1966, Soviet guidance, control, and computing were well behind the West's, and even their largest rockets were not up to the task of a lunar landing. Their early space capsules, the Vostok and Voskhod, were capable but primitive in comparison to Gemini and Apollo. Command and control on the ground was adequate, but not built up in the way that the US program was as soon as the quest to reach the moon was identified. Nonetheless, the USSR had a chance for glory. While a first landing on and return from the moon was doubtful, the possibility of sending a single, manned spacecraft in a figure eight loop around the moon was enticing. This was a far less ambitious goal than landing a man on the surface and bringing him home, but it would have profoundly reduced the public relations impact of the Apollo landings, and the Soviets knew it. This became a goal of their program as their own lunar landing efforts stalled.

In the US, the intelligence apparatus of the federal government worked diligently to unravel the secrets of the Soviet moon program. The newspaper *Pravda* rarely announced Soviet space intentions in advance, trumpeting successes only after they had occurred. Given this information gap and the early Soviet lead in space—the first satellite, the first man in orbit, and the first spacewalk—understanding their current capabilities and intentions seemed critical to winning the space race.

The NSA, CIA, and several think tanks provided their best estimates to the executive branch. Excerpts from these reports are telling. A 1965 National Intelligence Estimate, the reports provided to the US president and his staff, indicates a broadly correct assessment:

"The propaganda value of a manned circumlunar flight, and its simplicity and low cost relative to a manned lunar landing, lead us to consider this as a prime Soviet goal. If the USSR is not seeking to beat the US in a manned lunar landing, this project would probably be timed to precede the Apollo mission in an attempt to detract from the US achievement and to identify the USSR with manned exploration of the moon. Use of the present space booster for the circumlunar mission is possible but unlikely."[1]

To accomplish a lunar landing, the Soviets had instituted a crash program to build their answer to the American Saturn V moon rocket, the N-1. The N-1 had thirty engines on the first stage alone, as opposed to the Saturn V's five larger engines, and was extremely complex. It was underfunded, and its development and testing were behind schedule, when the chief designer of the Soviet space program, Sergei Korolev, died in 1966 (as mentioned in chapter 14). By then the N-1 program had started down a long path to failure. But a loop around the moon with a single spacecraft, as opposed to a tricky and complex landing, might be accomplished with already available technologies. This "flexible approach" was also noted in the US intelligence documents:

Some ambivalence in public statements by Soviet leaders suggests that they may be trying to keep their options open, but during the past year or so they have shown increasing caution, implying that the USSR has not in fact entered a lunar race. Thus, Khrushchev voiced both deep concern about the technical difficulties of such an undertaking and a willingness to profit from US experience and possible US failures. He and others expressed concern over the high cost of undertaking a manned lunar landing. His remarks in the past year clearly were intended to convey the impression that the USSR was not competing with the US in any race to the moon and to lay the foundation in the minds of the Soviet people that the US might be first. The new Soviet leaders have made public statements in a similar vein. After the flight of Voskhod, Brezhnev stated: "We Soviet people do not regard our space research as an end in itself, as a kind of race. . . . The spirit of frantic gamblers is alien to us." At the Kremlin reception for the Voskhod cosmonauts, Kosygin pointed out that the economic needs of earthly projects must not be forgotten in the rush into space.[2]

Despite this, there was a legitimate concern within the US intelligence community that the USSR might attempt a trip around the moon, thus muting much of Apollo's impact on world opinion. America might land there, but the Soviet Union would have *reached* it first. "There are no specific indications of Soviet intent to carry out a manned circumlunar project, but its relative simplicity compared to the manned landing mission, as well as its propaganda value as a major 'first,' lead us to consider it (along with earth-orbiting space stations) as a prime Soviet goal. If the USSR is not seeking to beat the US in a manned lunar landing, this project probably will be timed to precede the Apollo mission in an attempt to detract from the US achievement and to identify the USSR with manned exploration on the moon."[3]

The report concludes: "Soviet statements relating to the manned lunar mission can be traced back to 1961, when President Kennedy challenged the USSR to a space race with this as the specific goal. In considering how to respond to the US challenge, the Soviets would have had to assess carefully the benefits from such a project against those to be derived from other uses—military, space, and civilian—of the same resources. Equally important would be the Soviet leaders' view of their ability to compete successfully and their assessment of the consequences for Soviet prestige should disaster result from a project whose timing was dictated by the US."

So, during the frantic buildup of the US lunar landing program from 1961–1968, Soviet intentions were murky. As it turns out, the USSR was in fact keeping its options open, jumping from the relatively incapable orbital flights of Vostok and Voskhod to the lunar-intended Soyuz spacecraft. For Soyuz, two main tracks were being pursued: a lunar flyby and a lunar landing. There was a highly secret Russian lunar lander being designed, capable of depositing a single cosmonaut onto the moon's surface and bringing him back to the orbiting Soyuz (the Apollo lunar module carried two crew members). And a smaller booster and lighter hardware, usable for a flyby, was on a parallel track. Soyuz was modular, allowing for a number of flight configurations.

The Soyuz variant for the lunar flyby mission was referred to internally as the Soyuz 7K-L1 (the lunar landing mission version was the Soyuz 7K-LOK). While the massive N-1 would be needed to loft the mass needed for a landing, the lunar flyby mission could utilize the smaller Proton booster, already nearing completion and already a far less challenging project than the N-1. To make things more confusing, both versions of the Soyuz spacecraft were under the auspices of Korolev's design bureau, but the Proton rocket project was being run out of another agency under Vladimir Chelomei.

Both Soyuz designs were based on the earlier model that was first flown as Soyuz 1, the 7K-OK. The lunar flyby version had a larger antenna for communicating from lunar distances and was lightened to accommodate the long journey. The "orbital module" was also removed—the Soyuz had one module to accommodate the crew while in orbit, and another, separate descent module. The removal of the orbital module saved weight but cut the interior space roughly in half. It would be capable of supporting two cosmonauts for the flight in relative comfort (it was roughly one-third the volume of the Apollo capsule). Finally, the lunar flyby Soyuz would sport a toroidal (donut-shaped) fuel tank near its aft end to accommodate the extra fuel needed for the journey.

The Proton booster had its own teething troubles. First flown in 1965, Chelomei's rocket was initially intended as a super ICBM, for launching very large hydrogen bombs. Like the US, the USSR rocket program, in particular its ICBM designs, had moved from an early dependence on kerosene fuel with a "cold" oxidizer, liquid oxygen, to hypergolic fuels that could be stored at room temperature, far superior for the rapid preparation and launch of a nuclear missile. America's version of this propulsion system was the Titan, and Russia's was the Proton. Both used a form of hydrazine and nitrogen tetroxide—highly explosive and extremely toxic chemicals—as propellants. Korolev did not approve of these fuels for his rockets—he felt they were too dangerous—but with his death, Chelomei's rocket got a second chance. And it needed that chance and many more—after its maiden flight test in 1965, the rocket flew, and failed, countless times, not gaining final, full state approval until 1977. Nonetheless, with the N-1 moon rocket lagging, the Proton—fully approved or not—was the best bet for some kind of lunar spectacular. It was the only rocket powerful enough to send a Soyuz on a lunar flyby.

As the N-1 booster struggled to make it to the flight-testing phase—the Soviet landing program was now at least two to three years behind the Saturn V in terms of development—the circumlunar version of the Soyuz also needed to be tested. In a move that confused Soviet launch nomenclature even more than it already was (which is saying something), the test flights of the circumlunar Soyuz were called either Kosmos, for Earth-orbiting flights, or Zond, Russian for "probe," for lunar flights. Both appellations had been used previously: Kosmos was a broadly applied name for experimental flights, and Zond had first appeared in 1964 as an unmanned Mars probe that missed its launch date and was sent instead on a trajectory that took it past the moon—hence the more generic "Zond" instead of the more proper "Mars 2" or "Mars 3" of the era.

The Soviets had high hopes for the circumlunar program to salvage at least some of the prestige that was slipping from their grasp as their lunar landing program fell further and further behind schedule. Testing began in 1967, and continued through 1970. By the time the bugs were worked out of the circumlunar spacecraft, the US had landed four astronauts on the moon and the Soviets canceled their lunar flyby efforts—it would have been too little, too late. The N-1 lunar landing program quietly died soon thereafter. The official party line was in essence: we never wanted to go to the moon anyway; we're working on space stations. The latter part was true, but the former was not.

It's worth noting that the Saturn V rocket flew only twice before men were launched on it for the flight of Apollo 8, and the lunar module was only test flown twice before being sent to the moon. None of these tests were without fault, but none of the issues were showstoppers, and NASA's top management forged ahead with the program. In stark contrast, the Soviets flew the unmanned Zond version of Soyuz twelve times, never with complete success, and in the end canceled the lunar program outright.

The Zond was tiny when compared to the Apollo system: the Proton rocket was only capable of lifting about 50,000 pounds to orbit, while the Saturn V could lift approximately 210,000 pounds. And the Soyuz was smaller than the Apollo capsule and service module—5,500 pounds to Apollo's combined 32,400 pounds. But the Zond/Proton combination was the only chance the Soviets had to reach the moon.

The first two flights were into Earth orbit, so they were called Kosmos 146 and 154, and flew in March and April 1967. Kosmos 146 was a success, but 154 failed when it was unable to fire its upper stage.

Next up were the Zond flights. Two were launched in 1967, with both failing. In March 1968, Zond 4 was launched in a simulated lunar trajectory, but it was actually sent *away* from the moon—out to an equivalent distance—to avoid the complications of dealing with lunar gravitational effects. The flight seemed to be a success until reentry, when it lost control and headed for a landing outside of Soviet-controlled territory. The spacecraft was destroyed by a remote command during reentry.

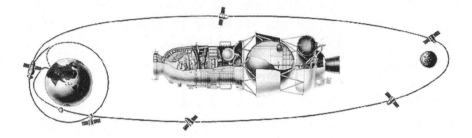

Fig. 15.1. Illustration of the design of the Zond capsule and its flight path for the Zond 5 mission. Zond 5 successfully flew a single loop around the moon with living passengers—microbes, reptiles, and insects. The spacecraft itself was based on the Soyuz but heavily altered to be launched on smaller rocket boosters than the ailing N-1. Image from NASA.

With Zond 5 the Soviets hit pay dirt, and came as close to sending living things to loop the moon as they ever would. There were a couple of false starts after Zond 4—a launch attempt the following month, in April, resulted in a misfired second stage, and in July another Soyuz test flight exploded on the pad, killing three technicians. But finally, in September 1968, Zond 5 rocketed into trans-lunar space, circling the moon two days later at a closest approach of about 1,200 miles. It did not enter lunar orbit; there were no rocket engines onboard sufficiently powerful to slow it into orbit around the moon, nor to break it free from orbit for a return to Earth. Those capabilities would remain the domain of the much larger, heavier, and far more capable Apollo spacecraft. Zond, aka Soyuz 7K-L1, was a stripped-down hot rod of a moonship, and would simply use the gravity of the moon to sling it back onto a return course to Earth.

There was a short-lived but potentially shocking moment in the flight, when radio signals with *voices* were heard coming from the spacecraft. Western observers had assumed that the flight was an unmanned test, but due to the traditionally tight-lipped Soviet media, this was an assumption only—the Soviets were not even admitting that this was a lunar mission yet. Late on September 19, a British radio telescope, the Jodrell Bank Observatory, heard human voices coming from the spacecraft. These were later identified as a recording, used to test communication from lunar distances, but it must have been a huge surprise for Western observers at first.[4] In any case, it was a sobering reminder that the USSR was close to sending humans around the moon.

The return to Earth was relatively uneventful right up until reentry. A malfunction in the onboard gyroscope (some sources identify it as a case of operator error) caused the reentry angle to be off-axis. For spacecraft returning from the moon at speeds of 25,000 mph, this angle is critical. The ideal alignment—the one the Apollo flights used, and the one the Soviets were aiming for—is a glancing path that allows the spacecraft to scrub off speed and energy when first encountering the upper atmosphere and prior to plunging into denser air. This was something the Zond flights attempted to do, but missed the mark again and again. Instead, the spacecraft went directly into a ballistic trajectory, coming home like a cannon ball. Besides intense heating, the Soyuz was subjected to over sixteen times normal Earth gravity—a force that the human body can endure for a brief period, but that would have likely been injurious or possibly fatal to any cosmonauts aboard. The capsule splashed down into the Indian Ocean, a novelty for the Soviet space program, which aimed for dry-land recoveries in Soviet-controlled territories.

The bobbing capsule, apparently none the worse for wear, was retrieved by two Russian ships deployed to the Indian Ocean. These ships were being closely shadowed by a US destroyer escort that watched the operation carefully. Zond 5 was hoisted aboard a Soviet ship and taken back to the USSR for investigation.

And what of the turtlenauts? As it turns out, while no humans were aboard Zond 5, there were a number of other creatures that could have qualified for some kind of flight medal from the Soviet leadership. Along with various plants and seeds, bacteria, some flies, and a nest of meal worms, two tortoises had been sent to the moon. While the tortoises had lost about 10 percent of their preflight body weight, they appeared to be none the worse for the journey, as were the insects. Like the early flights of the American Mercury spacecraft, which flew chimpanzees into space to check for any harmful effects, the Soviets had sent an odd cadre of organisms to the moon as a test prior to manned flights.

The Zond program continued to test the Soyuz spacecraft for a lunar flyby mission, including Zond 6 in November 1968. The final test flight was Zond 8 in October 1970, with two more tests canceled—the fire had gone out of the program after December 1968, when NASA's Apollo 8 spacecraft flew into lunar orbit on Christmas Eve. The novelty and public relations value of a Soviet flyby of the moon had vanished, and soon the Russians' entire lunar effort would be but a footnote in history.

There is an interesting postscript to the Zond story. In October 1968, Apollo 7 carried three astronauts into Earth orbit for ten days. The flight went well, and the stage was set for Apollo 8 to fly in early 1969. Apollo 8 had originally been intended as a test of the entire Apollo lunar landing system—the command/service module, the lunar module, and the Saturn V—in Earth orbit. However, the lunar module was far behind schedule, and while it had been tested in unmanned flight, the LM was not deemed ready to fly with humans aboard. But with a Saturn V and command/service module ready to go, it seemed wasteful to some in NASA's upper management to wait until the new year to fly the mission. A daring decision was made in August to send the spacecraft—sans lunar module—into orbit around the moon. The assigned crew quickly assented, as Apollo 8 commander Frank Borman wrote in his 1988 autobiography. He recounted being called into the office of Deke Slayton, who was in charge of astronaut selection, for an urgent meeting. "We just got word from the CIA that the Russians are planning a lunar fly-by before the end of the year," he recalled Slayton saying. "We want to change Apollo 8 from an Earth orbital to a lunar orbital flight. I know that doesn't give us much time, so I have to ask you: Do you want to do it or not?" Borman said yes immediately, only later running the idea past his crewmates (who, fortunately, agreed).[5]

This change in plans was kept a secret until November, with the crew quietly compressing a greatly revised training schedule into a few months. The mission launched on December 21, and reached lunar orbit on December 24, braking into a stable orbit around the moon with the command/service module's powerful rocket engine. After ten orbits, the engine was fired again to bring the astronauts home.

There has been much speculation since then about whether the US's intelligence agencies' reports about Zond 4 and 5 might have caused NASA to accelerate the push to reach lunar orbit in 1968. At the time, such queries were publicly denied outright. It was claimed to simply be the next logical step given the availability of flight-ready hardware and the desire to test it in the lunar orbital environment. Nonetheless, for decades, historians surmised that the Zond flights were a contributing or supporting factor, but not a decisive one.

Documents declassified in 2003 seem to be at odds with the official account. It appears that the Soviets were not the only ones with secrets. The CIA's deputy director for Science and Technology, Carl Duckett, was quoted as saying, "The likelihood that the U.S. will conduct a manned circumlunar flight with the Apollo-8 vehicle in December is a result of the direct intelligence support that FMSAC has provided to NASA on present and future Soviet plans in space." (The FMSAC was the CIA's Foreign Missile and Space Analysis Center).[6]

So, in the final analysis, the victory laps Apollo 8 flew around the moon may, in the end, have been prompted by a secret Soviet hardware test and a couple of turtlenauts after all.

CHAPTER 16

FALLING TO EARTH: THE DANGEROUS SCIENCE OF REENTRY

Spaceflight is dangerous. Space is a hostile environment—a body-rupturing vacuum, broiling heat, flash-freezing cold, destructive micrometeoroids, deadly radiation, and more. But the two most dangerous things about spaceflight are getting there and coming back. And for the last of these, the two prominent players in the space age—the US and USSR—had a variety of ideas of how they might attempt to rescue astronauts stranded in orbit.

The dynamics of reentry are fairly simple. An object in orbit around the Earth at lower altitudes, say between 100 and 1,200 miles, will eventually slow down and fall out of that orbit. This range is called low Earth orbit, or LEO. The lower the orbit the faster it will decay. Various forces are at work here, but the primary one is atmospheric drag, the same force that brought Skylab crashing to Earth years before NASA had thought it would.

Manned spacecraft in orbit around Earth operate in LEO, but do not remain there long enough to suffer orbital decay. More permanent structures, such as the International Space Station, will eventually reenter the atmosphere if not periodically boosted back to a higher orbit.

For spacecraft designed to travel from the Earth to orbit and back, reentry is initiated by firing rockets to slow the craft. As it loses energy, it begins to fall back to Earth. Once it enters the atmosphere, the friction caused by gas molecules further slows the spacecraft, and then later parachutes and sometimes braking rockets are used to slow it further, until it can land (or splash down) at survivable speeds. Space shuttles were slowed by atmospheric friction against the blunt under-body, then glided with wings to a high-speed landing on wheels. Space capsules are slowed by the friction against their blunt bottom, glide a bit laterally, then parachute down to the surface.

There are many dangers associated with reentry. First, the space-craft must be aligned properly when it encounters the atmosphere. This affects both where it will ultimately land, as well as its reentry trajectory—too steep an angle and it will suffer high-g loads, resulting in a crew that is either very uncomfortable or even seriously injured. If the angle is too shallow, the spacecraft can skip off the atmosphere and return to a lower orbit. In the case of the high-speed returns of the Apollo lunar missions, about 25,000 mph, a skip could have resulted in the capsule not returning to Earth for months. Earth-orbiting space-craft begin reentry at a speed of about 17,500 mph.

Second, the spacecraft must be protected from the intense heat caused by the friction encountered as it plunges through the atmo-sphere. The air simply cannot get out of the way fast enough, and the result is heating—anywhere from 2,000 to 2,600°F. The space shuttle experienced maximum heating of about 2,700°F. The Apollo capsules returning from lunar flights experienced heating of up to 5,000°F, due to their higher return speeds. The shuttle used a combination of heat-resistant ceramic tiles, carbon composites, and high-temper-ature fabrics to moderate the effects of reentry heating. Space cap-sules such as Mercury, Gemini, Vostok, Soyuz, the Chinese Shenzhou, Apollo, and NASA's new Orion, as well as SpaceX's Dragon, all use what is called an ablative heatshield. This is a resin-impregnated dish that resists heating both by shape—its blunt form causes a shock wave that diverts some of the heat away—and heat removal: the material actu-ally chars and erodes, and the ash that is carried away takes some of the heat energy with it.

Of course, the final step in returning to Earth is to land slowly enough so that the crew can survive. The shuttle touched down on its runway at a speed of about 220 mph, then rolled to a stop, shedding the last of its reentry energy as it traveled down the runway. Capsules must land vertically, so they need to be traveling not much faster than about 20–30 mph if landing in water; the slower the better. These kinds of spacecraft use parachutes, and, in the case of the Russian Soyuz and its Chinese variant Shenzhou, braking rockets are also fired just prior to encountering dry land.

As you can see, there are many ways to die during reentry if things don't work out as planned. Yet, in the long history of spaceflight, there have been remarkably few fatal reentry failures. The notable ones are the space shuttle *Columbia*, which disintegrated during reentry in 2003 due to damage to its wing heatshield, and the parachute failure on the Soyuz 1 flight.[1] Otherwise, despite parachute problems on capsules and missing tiles on space shuttles, most spacecraft have landed successfully with their crews none the worse for wear.

But the first step in returning home from Earth orbit is slowing the spacecraft enough to execute a controllable reentry. If the retrorockets don't fire when commanded, the spacecraft can be stranded in orbit for far longer than its crew could survive. This is a potential problem that has kept many engineers and managers up late at night—there is generally just one set of retrorockets, and if they fail, the crew could be marooned.

Over the decades since the first fledgling flights into orbit, a variety of rescue systems have been proposed, from spare rockets standing by at a launchpad to emergency bailout systems and safe-haven space stations. After the loss of *Columbia* in 2003, when the space shuttle flew high-altitude orbits such as the missions to repair the Hubble Space Telescope, another shuttle was placed on standby for a quick launch if needed. Other scenarios (with lower orbits) dictated that a shuttle disabled in orbit would simply return to the space station and dock until another spacecraft could be sent to rescue the crew.

In the early days of the space age, prior to Salyut and Skylab, no such contingencies were available, but there were a number of plans for rescue craft. As early as 1962, designs were drafted based on the flexible Gemini spacecraft for rescue operations near and far. One plan, called Rescue Gemini, would add a lengthened back structure—ten feet longer than a standard spacecraft—to provide just enough space for three additional astronauts (for a total crew-carrying ability of five), and a maneuvering system.[2] Upon reaching orbit, the pilot would fire up the rocket engines in the back of the spacecraft, match orbits with the stricken capsule, and the astronauts would EVA to safety in the Rescue Gemini for the ride home. It was a 30,000-pound beast, about five times the weight of the regular Gemini, and was never built.

Another Gemini variant, proposed in 1966, was for a Gemini lunar rescue vehicle. This was an outgrowth of various studies that had looked at the Gemini capsule as a possible lunar landing craft, but in the end it was decided that a Gemini lander would only shave a year or so off the Apollo program (which was already being constructed) and would be an unnecessary diversion of funding. But as a rescue craft, a similar idea would serve different goals. Launched on a Saturn V, the Gemini Lunar Surface Rescue Spacecraft would fly to the moon unmanned and under robotic control. The Gemini would be enlarged to accommodate up to three astronauts, the crew complement of an Apollo mission. One version of this design specified the use of either three Apollo lunar module descent stages to land on the moon, ascend from its surface, and return to Earth, or two service module propulsion units (the cylindrical part behind the Apollo command module) and a lunar module descent stage to provide the same functions. Yet another variant suggested a Gemini capsule affixed to an ascent stage atop a lunar module descent stage. This version would be pre-positioned on the moon for rescue of astronauts stranded by a malfunctioning lunar module, deemed the most likely failure scenario of the Apollo missions. These studies remained on paper only.

Most spaceflight operations are confined to Earth orbit, however, and options for rescue from LEO were considered as early as the late 1950s when the US Air Force was planning the X-20 Dyna-Soar spaceplane. During the X-15 program, there was an increasing concern that standard bailout systems—ejection seats with parachutes—would not be effective at the very edge of space, and would obviously not work from orbit. An early proposal for a very high altitude rescue system was called the Paracone. This was constructed of René 41, the same metal alloy used to construct the Mercury and Gemini capsules and much of the X-15. The flexible metal disk would be attached to the back of an ejection seat, and when the pilot bailed out of the X-20, it would expand into a round concave shape reminiscent of a huge contact lens. Upon the firing of retrorockets, the Paracone would slow enough to reenter the atmosphere and descend to a (relatively) soft landing under parachutes, with the final bits of energy absorbed by a crumple-zone structure on the back of the Paracone. It would have been a wild ride to say the least.

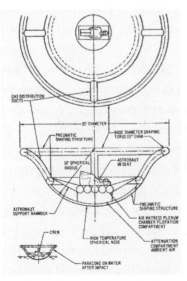

Fig. 16.1. The Paracone was an early alternative for aircraft that flew too high to use traditional ejection seats. Its effectiveness was never tested. Image from NASA/USAF.

Another proposed emergency return system, perhaps the wildest of all the design proposals, was called MOOSE, which stood for Manned Orbital Operations Safety Equipment (although at one point in its development it was reportedly an acronym for Man out of Space Easiest). MOOSE was intended to bring one astronaut back to Earth from any malfunctioning spacecraft that was capable of carrying the suitcase-sized, 200-pound rescue unit. It was an ingenious design by any measure. When all options for returning in the spacecraft were exhausted, the space-suited astronaut would open the hatch and exit with the MOOSE unit. It was in essence a big plastic bag with a heat-shield. After the bag was unfolded, the astronaut would climb inside, lying in it as if it were a six-foot-diameter beanbag chair with a thin plastic cover. The bag had a one-fourth-inch-thick heatshield on the back—due to the low mass of the system (less than 400 pounds with an astronaut inside), the heavy heatshield of a space capsule was not needed. Once inside the bag, the inflation system would be activated—two pressurized canisters would fill the back half of the disk-shaped MOOSE with expandable polyurethane foam. The foam would stiffen instantly and form a roughly conical dish. This gave the MOOSE some structural rigidity, and would help insulate the astronaut from the heat of reentry and impact of landing. When preparations were complete, the astronaut would fire a small rocket pack that extended out from the front of his suit, slowing the MOOSE sufficiently to reenter the atmosphere. It was designed to be stable during reentry, keeping the heatshield oriented generally toward the ground. At about 30,000 feet a single parachute would be released, and the MOOSE would float to a soft landing in water, in which it could float just like a life raft. Brilliant as it was, MOOSE would have been a risky, last-ditch option and was never developed beyond very basic ground testing.

These ideas have been mothballed for decades, just another set of wild space-age designs that never made it off the drawing board or beyond limited prototyping. But there is new life in some of these old concepts. With Richard Branson's Virgin Galactic and Amazon's Jeff Bezos's Blue Origin looking to haul tourists into space, a handful of companies have begun investigating high-tech space suits that would allow an adrenaline-fueled few to exit a spacecraft and descend to Earth protected by nothing more than the suit. In 2013, a high-altitude suit called the RL Mark VI Space Diving Ensemble was announced by a pair of companies working together out of Baltimore, Maryland. It is an advanced pressure suit, complete with gyroscopically stabilized boots to keep the user in the proper orientation during free fall and reorient him or her to a feet-down position for final deceleration and landing. Current designs do not use a parachute—the suit is intended to slow the user via its aerodynamic design and, believe it or not, Iron Man–like rocket boots. The suit would be insulated from the atmosphere by aerogel and temperature resistant Nomex cloth similar to what was used on the upper surfaces of the space shuttle (Nomex is also used by race car drivers to protect them from fire in case of a crash). Jumps from true low Earth orbit will likely require more advanced protection systems—current designs cap out at sixty-two miles, the widely accepted boundary of space.

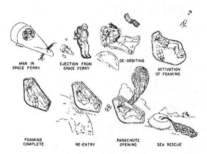

Fig. 16.2. MOOSE was a more compact unit than the Paracone, and had to be manually unpacked and inflated for use. It could, however, be stored in a much smaller space. Image from NASA/USAF.

Suits such as the RL Mark VI are "in development," so don't expect to see anyone diving from the edge of space anytime soon. However, record-setting high-altitude balloon jumps by Felix Baumgartner and others at altitudes in excess of 24,000 feet have shown that humans in pressure suits can survive free fall from extreme heights—suits like the RL Mark VI merely add a higher altitude (and possibly orbital reentry) component to their wish list, requiring protection from heating. Such products would be intended primarily for extreme recreation, but could also solve the problem of emergency reentry from stricken spacecraft if properly engineered.

There's one last installment to these tales of unusual reentries, one not intended to assure the survival of an astronaut—or, in this case, cosmonaut. In 2006, a worn-out Russian space suit was repurposed as a short-term satellite and dubbed SuitSat. The retired Orlan suit was modified with sensors and a radio transmitter—which was inelegantly mounted to the top of the helmet—and flown to the International Space Station. On February 3, 2006, the modified suit was stuffed with cloth to give it some rigidity, and the electronics were activated (a radio transmitter beamed recordings of greetings collected from children worldwide) and checked for proper function. It was then escorted out of the ISS by astronaut Bill McArthur and cosmonaut Valeri Tokarev during a scheduled EVA. Nicknamed "Ivan Ivanovich," when the suit was released, Tokarev said, somewhat cryptically, "Goodbye, Mr. Smith!"[3] The voice signal and telemetry of suit temperature and system condition was transmitted as the faux cosmonaut drifted slowly away from the space station—an odd and chilling sight.

The suit transmitted for a short time with ground stations reporting limited success in receiving the signal—amateur radio enthusiasts had supported the experiment with relish. It reentered the atmosphere on September 7, seven months after it was pushed into oblivion, burning up over Australia (just as Skylab did, but much more quickly).

The SuitSat experiment was reminiscent of a Ray Bradbury short story, "Kaleidoscope," which appeared in his anthology *The Illustrated Man* in 1951.[4] In it, a group of spacemen (the term "astronaut" was not in use yet) find themselves drifting in space after their rocket is damaged and they are unexpectedly expelled from the torn hull. As the men drift apart, they talk to each other over the radio, expressing dark humor, memories, and regrets as they slowly move out of range of one another. At the end of the story, the last man we hear from, named Hollis, is descending into the atmosphere, wondering if anyone will see him as his life ends in an incandescent flash. The scene changes to the Earth, where a young boy points to the starry nighttime sky and says, "Look, mom, look! A falling star!" She replies, laconically, "Make a wish. . . ."

CHAPTER 17

FUNERAL FOR A VIKING: THE END OF VIKING 1

UNCLASSIFIED

t just took a moment. A few seconds made the difference between the continuation of NASA's longest mission success to date, and the end of a program. As a thin breeze blew across the dusty wastes of Chryse Planitia, Greek for "Golden Plain," near the Martian equator, a small three-legged lander began to stir. Covered in a thin film of fine red dust, Viking 1, America's first machine on Mars, had not been active for some time. Its manipulator arm had sat in the same position for many months, and nothing on the outside moved. Only the slow, heat-producing decay of the nuclear plutonium fuel and the heat coming from the electronics within the lander would give an observer any idea that it was functional. Viking 1 would have seemed inert—it could have been a sculptor's homage to a golden age of planetary exploration.

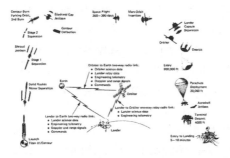

Fig. 17.1. A profile of the Viking mission. From the point that the lander detached from the orbiter, top right, and began its descent toward the Martian surface, it was on its own—radio signals took about eighteen minutes to reach Mars one way. The primitive onboard computer would guide the lander to the selected landing zone, and all the controllers at JPL could do was wait for a confirmation of a successful landing. Image from NASA.

Then a small whirring noise indicated activity within the lander—a mechanical tape drive, the memory storage for the small onboard computer, was recording commands being sent from Earth, barely visible as a small blue dot in the salmon-colored sky. New programming instructions from home were recorded and executed. Within a short time, the blue dot moved on in the heavens, and while not seemingly noteworthy at the moment—it happened every twenty-four hours, forty minutes—that was the last radio conversation between Viking 1 and its masters. A magnificent mission of exploration and discovery had come to a sudden and unscheduled end.

Back on Earth, at NASA's Jet Propulsion Laboratory (JPL), perplexed engineers were soon poring over data printouts. The small team that had been charged with the continued monitoring of the last of four Viking spacecraft on and around Mars realized something was wrong—there was no radio signal coming back from the lander. It should have acknowledged receipt of the new program they had sent up, and continued to send back its regular reports—temperature, wind speed, and other data continually gathered since its arrival in July 1976. But there was no downlink, no signal at all. The lander had gone quiet, and the team began to search for clues as to what could be causing the interruption. Given the age of the machine—it had left Earth atop a Titan III rocket in 1975—and the complexity of the computers and electronics aboard, it could be a number of things. It took time, but working closely with the Denver-based builder of the lander, the Martin Marietta Corporation, they soon found the cause. Over the next few days, the ramifications sank in: they had a serious, and potentially mission-ending, problem.

The challenge to explore the surface of Mars had a long gestation period, reaching back into the first years of NASA. Early plans included a program called Voyager—which had no relation to the Voyager missions to the outer planets—that would have used a Saturn V rocket to deliver a large and heavy payload to the surface of Mars, possibly something derived from Apollo flight hardware. But such a massive lander would need to slow down enough once it entered the Martian atmosphere to survive touchdown, and as the scientific results came back from the early Mars probes—NASA's Mariner 4, 6, and 7 flyby missions—it became clear that the atmosphere was much, much thinner than originally thought, and such a huge machine would not be able to slow down enough to land softly. So the design was changed to a smaller lander and separate orbiter, derived from the Mariner probes (for the orbiter) and the Surveyor moon lander (for the lander). This combination was still slated to be launched on a Saturn V. The program eventually morphed into the Viking program, which retained the idea of separate orbiters and landers, but were launched on the cheaper Titan III rocket. The Viking design was streamlined—the original plans called for something the size of an SUV to be landed on Mars, but the designers had to downsize the spacecraft, and make it much lighter, to have any hope of success.

The program moved ahead in this "de-scoped" design with NASA's Langley Research Center working with JPL, and the Martin Marietta Corporation building the orbiters and landers. The Viking project was truly a leap into the unknown. At the time, the Mariner missions had returned data from Mars, but much about the planet was still a mystery. However, with the first experience of landing a robotic probe on the moon with the Surveyor 1 starting in 1966, and flyby successes with the Mariner spacecraft at Mars and Venus, there was at least a technological base on which to build. It was just a base, though, and the project to send two sophisticated machines to orbit Mars and two more to land on its surface was a vast leap beyond what NASA had accomplished before.

It's worth noting that at the time robotics were primitive. Two identical spacecraft were generally dispatched for any planetary mission, one backing up the other. Viking was no exception. The twin orbiter/landers were a huge commitment at one billion 1970s dollars in all (or about $6.2 billion today).[1] What were the agency's motivations and goals?

In NASA's own words:

> Recognizing that many scientific secrets still lie hidden throughout the solar system, NASA has a program of solar system exploration

aimed at answering the following questions: (1) How did our solar system form and evolve? (2) How did life originate and evolve? (3) What are the processes that shape our terrestrial environment?

Our astronauts have begun detailed exploration of the Moon, but we have sent only a few instrumented spacecraft past or into orbit around the other planets. Among the other planets, Mars is the most potentially rewarding as an astronautical objective, especially in terms of the second question. It is neither too hot nor too cold; it possesses carbon dioxide and some water. Life could exist there, and scientists are eager to send their instruments down to the Martian surface. The possibility of Martian life—extinct, extant, or future—is the target of the Viking program that is described in detail in this publication. The two Viking spacecraft, to be launched in 1975, will be Orbiter-Lander combinations. The Orbiters will contribute to the science objectives of the mission by taking photographs and spectra over large regions of the planet. The Landers will make in-situ atmospheric and meteorological measurements during descent and while on the surface. Once safely landed, various other instruments will analyze the soil for organic and inorganic compounds and try to detect biological activity. Viking is a challenging program to explore the surface of a planet millions of miles away. From the information in the stream of radio signals beamed back to Earth across that immense void, we hope to learn more about Earth through the study of the differences and similarities of the planets and, possibly, to hear first signals announcing the discovery of extraterrestrial life.[2]

This search for life, to some scientists, involved a bit of scientific hubris. At least one of the key scientists on the mission, Norman Horowitz of Caltech, felt that many behind the mission were clinging unrealistically to the ideas of Percival Lowell, that they were die-hard romantics. He felt it very unlikely that life would be found beyond Earth, certainly not something we would recognize. He also questioned the assumption that life would be found as we are used to seeing it, with water as a key element. He knew that liquid water was unlikely to be pooled anywhere on the Martian surface, and questioned its use in the Viking life science experiments, as was planned.

As Horowitz put it years after the Viking program, "It's hard to convey in a few words the total commitment people had in those days [the early 1960s] to an Earth-like Mars. This was an inheritance from Percival Lowell. It's amazing: in pre–Sputnik 1 days, in fact, up till 1963, well into the space age, people were still confirming results that Lowell had obtained [through his telescope]; totally erroneous results. It [was] simply bizarre!"[3]

But in the end, the mission made compromises to encompass the attitudes of various members of the life science team. The challenge before them, especially for the lander, were large. JPL had gotten proficient at building orbiters, and had met with great success at Mars in 1971. But a Martian lander, which would need a new computer to accomplish its complex tasks, including the difficult landing, was a completely new undertaking. The space age was only about eight years old when the planning for this began. Computers like the IBM System/360 filled entire rooms, and only once before, for Apollo, had anyone designed such a small computer that could survive the rigors of spaceflight. In addition, the idea of a small life science lab, much less one that would survive the torturous flight to Mars, tumultuous landing, and savage surface conditions, seemed to be pure fantasy. The temperature swings on the surface alone, from a comparatively balmy 70°F to –225°F, were a potential showstopper.

The goals were broad and life-oriented:

> The goal of the NASA Viking program is to learn more about the planet Mars by direct measurements in its atmosphere and on its surface. . . . On both the Orbiter and the Lander the primary emphasis will be on biological, chemical, and environmental aspects of Mars which are relevant to the existence of life. This would be the main consideration. The Lander carries by far the most instruments. It is, in fact, a miniature automated laboratory.[4]

Since NASA's first discussions of landing a machine on Mars, the notion of searching for life had been a key goal. There was much debate, primarily over whether to orient the surface science investigations toward a search for water, or jumping ahead to look for microbial life in the soil. In what may now appear as an unreasonably sunny assessment of the planet, but with sound reasoning at the time, it was decided to seek signs of microbial activity in the Martian soil. It was a daring endeavor that would not be rewarded with a clear answer.

The question of how to best detect life on another world became its own debate between the life science researchers. Two instruments had been defined as essential early mission planning: a gas chromatograph and a mass spectrometer. Both should be able to detect the chemical elements in very small amounts of gas exuded by heated soil samples. This was the first deployment of such devices into space.

The arguments over the remaining experiment options centered on the notion of whether or not any microbes on the planet would metabolize nutrients similarly to Earth-based organisms. The experiment package ultimately included instruments that were thought to be capable of detecting microbial life in either case—whether it ingested and metabolized onboard nutrients (which would be squirted into the soil sample) or not. Three more experiments were chosen with an eye toward sensing microbes in the soil.

Teams of engineers and designers joined forces to shrink the needed hardware down to the required size and weight parameters. It would be a difficult task; these were traditionally large, heavy laboratory instruments, and making them portable enough to fly to Mars was a huge undertaking, but in the end it was done with great elegance. Identical instrument packages flew on the two landers.

To accomplish the many goals of the landings, from the science experiment package onboard each lander to the difficult descent scenarios the two machines faced, a proper, flight-rated computer had to be selected. It was a primitive affair by today's standards, a small, basic processor with 6k word memory—it would barely run a modern toaster oven. Even at the time it was not the most advanced unit available, having origins in the Minuteman ballistic missile program of the 1960s. The memory boards used magnetic cores—small, donut-shaped ferrite disks—with hair-thin wires run through them. If the wire went through the hole in the donut, it was one value; if it bypassed the donut, it was another. It was already old technology when the Vikings flew, but very robust and resistant to many of the dangers the spacecraft faced. This type of memory was hardwired (to change a value you had to disassemble the boards and reroute the wires), so for updatable instructions, a separate data drive was added. This was the tape recorder, acting like a linear and very slow hard drive, that would record and store updates to the landers' programming. It would become the central player in Viking 1's dramatic demise in about five years.

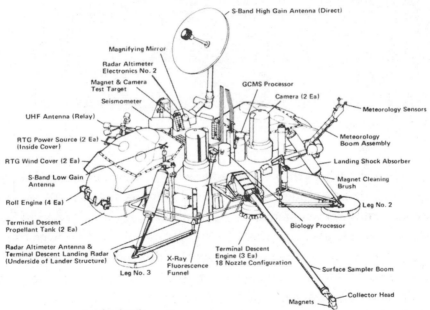

Lander details

Fig. 17.2. Schematic view of the Viking lander. At top is the main receiving dish that would cause so much consternation at the conclusion of the mission. The cameras are housed in the two units on the main deck that look like oversized soft drink cans. Image from NASA.

Fig. 17.3. A technician works carefully with a circuit board from the Viking lander's computer. Image from NASA.

The Vikings launched in August (Viking 1) and September (Viking 2) 1975. Viking 1 entered orbit around Mars on June 19, 1976, and Viking 2 followed on August 7. The combined orbiter/landers circled Mars while the planetary scientists studied new images received from the high-resolution cameras. The improved resolution brought new worries; a boulder smaller than a beach ball could kill the lander, and what they saw was not encouraging.

Prior to the arrival of the Vikings at Mars, the best tools mission planners had were images taken by the Mariner flybys, and, later, those returned by Mariner 9, which orbited Mars in 1971. Mariner 9 provided a comprehensive map of the Martian globe, and that map told a story, one of geologic upheaval in the distant past with vast flooding and erosion events, followed by eons of scouring by wind. There were areas that were clearly off-limits for a landing—the poles were too high in latitude and covered in ice, and many other regions were criss-crossed by dry valleys, boulder-strewn river deltas, and intense cratered basins. There were some places that looked like sandy desert highlands, however, and they promised to be smoother.

"Smoother" is a relative term when discussing landing targets in the 1970s. While Mariner 9 was able to map most of the planet, its very best image resolution was about 300 feet per pixel. This meant that anything smaller than a good-sized apartment building was going to be invisible, and this did not offer much comfort to planners trying to avoid beach ball–sized obstacles. They had to generalize their information, and infer from area-wide patterns—if a certain region appeared to be relatively feature-free, with little topographical disturbance, that might lessen the chance of setting the lander down on a large-ish rock. Might.

As one of the mission managers put it to the press when discussing the Mariner photos, "The visual images are only really telling us what is observable at . . . 100 meters and up. . . . [There are still] Rose Bowl size hazards."[5]

The Viking pictures were an order of magnitude better than those from Mariner 9, spotting objects as small as twenty-five feet across, so once again, the mission planners met to agonize over their choice of landing zones. The one they had picked based from Mariner 9's maps now looked like a potential disaster, with a rumpled, water-etched surface that had been completely invisible before. There were many other factors to consider: the geologists still wanted an interesting area to study, which meant looking for a site close to a major impact event (a crater) or a river delta with the resulting runoff and sediments. The life science people wanted something that might be conducive to their experiments. The embattled engineers just wanted something safe; something that looked like a huge, blank parking lot would suit them fine. They all knew that the lander would have to navigate down to the surface alone—Mars was so far away that a one-way radio message took eighteen minutes to cross the void; there would be no live-joysticking a landing from JPL with a round-trip of almost forty minutes for each reaction and corresponding command. All they could do was give the lander the clearest target they could, set the guidance system, and say good-bye. After that, it was on its own.

A month after the Vikings arrived at Mars, and after much agonizing and debate, the choices had been made and JPL committed to a landing site. Viking 1 headed down on July 20, 1976, landing in Chryse Planitia at 5:12 a.m. Mission Control time (PDT) on a flat, smooth area surrounded by boulders and smaller stones. By pure serendipity, the lander avoided contact with any of the hazards that surrounded it. After more debate, Viking 2 landed on September 3, halfway across the planet, coming to rest at a slight angle. It too survived the descent into rubble-strewn terrain with grace. Humanity now had not one but two robotic emissaries on the red planet, and the work of discovery was underway.

The orbiters continued their tireless journey around Mars, mapping the landscape below and focusing attention on regions of interest on the surface. The two landers made intensive investigations of the Martian atmosphere, weather, and soil. Within days of arriving they each scooped up soil samples with their robotic arms, deposited some of the Martian regolith into the onboard laboratories, and had run their analyses. The results were surprising . . . then shocking . . . and then ultimately discouraging. While early readings showed a strong reaction in one of the life science experiments—something appeared to be responding to the nutrient solutions being injected into the soil samples—follow-on experiments indicated that it was probably not a biological entity, but likely a chemical reaction of harsh chemicals in the dirt itself. The best guess was that Martian soil was infused with perchlorate, a highly reactive chemical that had reacted to the all-too-rare presence of water used in the experiment.

Other experiments were conducted, testing the dynamics of the soil—grain-size composition, slope holding, and so forth—but the big answers came in early in the mission. No positive, confirmed detections of life were received. Public attention drifted, but the mission continued with vast amounts of data and imagery coming in from both landers and orbiters. It was a spectacular demonstration of robotic exploration—in so many ways, the first of its kind—that set a high bar for those to follow.

And then, one by one, the machines began to fail. These were large and complicated spacecraft, operating right at the limit of robotic capability in the 1970s. Today we expect such missions to last well beyond their warranties—one Martian rover, Opportunity, is driving into its thirteenth year on the red planet as of January 2017. The Vikings also lasted well beyond their primary missions, intended to be a minimum of one year for the orbiters and ninety days for the landers.

The Viking 2 orbiter was the first to succumb after about two years of operations. It had developed a gas leak in its maneuvering fuel supply, and was shut down in July 1978, after 700-plus orbits of the planet. Its last act was to raise its orbit to avoid reentry into the Martian atmosphere and possible contamination of the planet's surface. The Viking 1 orbiter was shut down in August 1980, due also to having run out of maneuvering fuel, and it was elevated in a similar manner as its sibling to avoid planetary contamination.

The Viking 2 lander operated until April 1980 when its batteries failed. The lander's nuclear fuel supply would have operated for many more years, but lacking the onboard batteries' ability to store and regulate power, as well as keeping the spacecraft heated through the cold Martian nights, the lander was not able to continue functioning.

But it is the story of the Viking 1 lander that truly touches the heart. It continued operations well into 1982. Had its batteries held up, the plutonium-fueled power generator could have kept it going for a decade or more. But it was not to be.

Six years into its groundbreaking mission, the support staff had been reduced to the bare minimum to keep the spacecraft healthy and to handle the data it transmitted to JPL regularly. There were other planetary exploration programs to attend to—the Voyager probes to the outer solar system had launched in 1977 and, along with JPL's other responsibilities, demanded ever-increasing levels of staffing.

The small team minding the remaining lander monitored its

onboard systems continuously, and in late 1981 noticed a serious anomaly with its batteries. Having experienced the loss of the Viking 2 lander for similar reasons, they worked to quickly come up with a solution to the degrading power storage system. The team solicited input from experts all over the country, and settled on a new recharging routine, one that would charge the batteries to a known voltage instead of for a fixed time, as they had been doing since the mission began.

New software was written to accomplish this change and was uplinked to the lander. The controllers sent the command skyward on November 19, 1982—a date that they would remember for many years to come. They waited for the confirmation from the lander that it had received the uplinked instructions . . . and waited, and waited. No signal arrived. The team declared a spacecraft emergency the following day.

It was thought—and hoped—that a low-voltage switch had tripped due to the ailing batteries. Subsequent commands were sent to close the switch and increase the battery-recharge cycles. But that did not work, and before long they realized that a terrible mistake had occurred in the software upgrade.

Programmers digging through the computer code saw that the new instructions had inadvertently been placed on an off-limits part of the lander's data tape recorder memory. The commands to alter the battery-charging cycles had written over the program that repositioned the high-gain antenna dish to track the Earth. The antenna was no longer pointing properly, and would likely not do so again. The team scrambled to come up with a fix—it was the first time the term "tiger team" had been used since the Apollo 13 emergency, when an exploding oxygen tank required all hands on deck to bring the crew home alive. And while Viking was a robot, its minders felt as if the lander were a fellow explorer on Mars who was in trouble.

Within a week, the tiger team had designed a series of command uplinks that should, if the lander received them, result in renewed contact with Earth. Controllers were guardedly optimistic. But by the end of December, there had been no change—there was no signal from the lander, and the attempts were suspended.

These commands had been manually input for transmission, a time-consuming and potentially error-prone process. Working again with Martin Marietta, an automated system was designed to reproduce the manual command sequence. This was first transmitted in February 1983, and continued through the end of that month with no success. The recovery effort was permanently terminated at the end of that month.

It is not known how long Viking 1 functioned, staring at a bare patch of sky, empty except for faint stars. It may have been months; it could have been years. But eventually the first spacecraft on Mars ran out of battery capability, and if the low voltage switch had not been tripped before, it surely was after a long period of inactivity.

This ended the Viking program, the first successful landing on another planet and a spectacular mission of exploration and discovery. It would take over twenty years for a machine to return to the Martian surface, the Mars Pathfinder lander and rover that landed in 1997. That mission, too, was a rousing success, but it owed much of its achievements to the Vikings, the first emissaries to the surface of Mars.

CHAPTER 18

SAVING SKYLAB: COWBOYS IN SPACE

UNCLASSIFIED

As Gene Cernan, the last astronaut to stand on the moon in 1972, prepared to climb the ladder into the lunar module, he said these words:

> As I take man's last step from the surface, back home for some time to come—but we believe not too long into the future—I'd like to just [say] what I believe history will record. That America's challenge of today has forged man's destiny of tomorrow. And, as we leave the Moon at Taurus-Littrow, we leave as we came and, God willing, as we shall return: with peace and hope for all mankind. Godspeed the crew of Apollo 17.[1]

But we did not return to the moon. In fact, the magnificent infrastructure of the Apollo program was used for only five more flights, for two programs, then scrapped.[2] The Nixon administration made a decision to develop the space shuttle as a cheaper alternative to the Saturn V and Apollo spacecraft (the shuttle ended up being more expensive), and the three remaining flights of the Apollo program, Apollo 18, 19, and 20, were canceled. Most of the remaining moonships were hauled off to museums where they reside today, reminders of an amazing epoch of space exploration.

Of those final flights of Apollo hardware, Skylab was the more remarkable of the two (the other being the Apollo-Soyuz mission), and was the only major part of NASA's extensive post-lunar landing planning that was implemented. Flights to Mars, flybys of Venus, and many other daring programs and missions were studied ad nauseam, but the one that made the final slashing cuts of the new presidential administration was America's first space station.

While American astronauts were exploring on the moon, the Soviet Union was hurriedly building its first space station, called Salyut 1 (the military name for the station was Almaz). While publicly saying that they had never intended to race the US to the moon, the Soviets had experienced terrible setbacks and losses throughout the 1960s, working tirelessly to get a cosmonaut onto the lunar surface, or at least to loop the moon. But these efforts failed miserably, and the only thing they could do now was find a way to salvage their flagging prestige. Salyut was a fast-paced effort that launched in 1971. To the Soviets' credit, the space station was a success, though not without its trials. The first Soyuz spacecraft that tried to dock and occupy the station was not successful, and had to abort the mission and return home. The second mission, Soyuz 11, was initially successful, with the crew of three occupying the orbital outpost for twenty-three days. Unfortunately, the Soyuz capsule depressurized during reentry, killing the cosmonauts upon their return.[3] A few months later, low on fuel, Salyut 1 was deorbited and burned up upon reentry. More Salyuts would follow—it was what MOL should have been, a successful, moderately budgeted, small space station. But NASA felt it had a better plan.

Skylab had been in the works for some time. It was a product of the many, many studies of what could be accomplished using Apollo hardware after that program ended. The Apollo Applications Program (AAP) office was a small operation in NASA's vast empire, grinding out one paper after another in alignment with the Apollo Extension Series, founded to devise additional uses for the magnificent Saturn IB and Saturn V rockets and Apollo spacecraft. The production lines for the boosters had ground to a halt in 1966 as the final moon rockets were completed, and NASA wanted to find additional missions for the machines. Extended lunar surface missions, an Earth-to-moon shuttle called the "lunar taxi," and missions to Venus were considered (as discussed in chapter 7).

The AAP studies that led to Skylab involved repurposing an upper rocket stage as an orbiting outpost. Two approaches were studied, one using a spent third stage of a Saturn IB rocket, the S- IVB stage, once it had used all its fuel. This was called the "wet workshop." The upper stage would be fueled and launched into space, and when the fuel was exhausted, hydraulic pistons would push various structures out from the interior sides—workstations, instrumentation, and crew quarters that folded up neatly when the stage was full of 70,000 gallons of liquid hydrogen (a second tank, containing liquid oxygen, held another 20,000 gallons). The rest of the interior—items that did not do well immersed in cryogenic fuels—would have to be assembled by astronauts once the last of the fuel had been emptied from the rocket stage. It was an ingenious and audacious design, but not one without risks.

The second approach was called the "dry workshop," which would launch an empty S-IVB into orbit atop a Saturn V, already outfitted for use. This was a far safer approach with many fewer variables, since the completed space station would be assembled and rigorously tested on the ground prior to launch, and would fly without fuel. The station would be ready to use when astronauts arrived in a second spacecraft. Wisely, this design was selected to fly in 1973. But even with the increased safety and reliability factor, it was a tough project.

The program, while important, was not a top priority, as NASA was intently focused on accomplishing the first moon landings in 1969. But Skylab was well studied by the time the first astronauts set foot on the moon, because, among other reasons, it was doable. No new rockets were required, and no multiple launches and assembly (that process would come to fruition later, with the International Space Station). Additional hardware was limited to an airlock module, docking adapter, and a telescope system.

Other designs for a space station had floated around NASA since its inception—people had been discussing the possibility of a crewed orbital complex since the 1920s. This discussion was brought into the public consciousness when Wernher von Braun unveiled his plans for a huge military space station in the 1950s as a part of a unified program to explore the moon and Mars (see chapter 3). But even in a time during which flags were being jammed upright into lunar soil, a modular orbital outpost was still a fantasy. What was not a fantasy was the idea of converting a Saturn rocket's upper stage into a good-sized space station.

Skylab's core would measure forty-eight by twenty-two feet—the measurements were far larger if you included the solar panels, docking adapter, and airlock. The station would weigh almost 80,000 pounds, about double the mass of Salyut.[4] Two complete Skylab assemblies were completed, with only one flown.

Skylab was a masterpiece of repurposing existing hardware to fit new needs and objectives. To visit the remaining parts of the program on display—there are a few scattered at NASA-affiliated museums around the US—is to truly appreciate the scale of the project.[5] Other than the Salyut, the largest pressurized volume in space had been Apollo's combined command module/lunar module (when docked), with a total volume of a very small travel trailer at best. Skylab would have almost 13,000 cubic feet of interior volume—still the largest single spacecraft ever flown (the ISS is larger, but made up of multiple components), and the most voluminous single pressurized space. Stepping into a Skylab exhibit, you have a sense of the vastness inside. Then, imagine being weightless, and that interior space feels much larger.

The plan was for crews to spend up to a couple of months at a time in the space station, and for once, *form* was going to be important, not just function. Until this time, NASA's spacecraft had been practical and spartan, but nobody had spent more than two weeks in one. Taking human factors into account, the agency hired the famed industrial designer Raymond Loewy to do an interior makeover of the engineering mock-ups. Loewy had designed, among other things, the Studebaker Avanti sports car, the Greyhound Scenicruiser bus, stylish Coca-Cola vending machines, and endless other mechanical works of art coveted by the industrial age. He would now be responsible for the mental well-being of America's newest space heroes, the Skylab astronauts. Loewy designed the overall color scheme, the wardroom where the astronauts could eat and relax, and specified the placement of a large window from which to admire the Earth below. NASA was not thrilled with that final item when it was presented—windows were a risk deemed unnecessary by the engineers (smaller portholes would have been acceptable, perhaps), but the psychologists got a few words in and the station would have a reasonably sized window. It was a major contributor to the mental health of the Skylab crews, according to their postflight reports.[6]

McDonnell Douglas received the contract for Skylab in 1969 and completed the station in short order in 1972. The contract included a training module for ground use and a backup station. One surviving vestige of the "wet workshop" design was the open-grid aluminum flooring inside the station, originally designed to allow fuel to pass through as it was consumed during flight. As it turned out, the grids were still useful, allowing for attachment points for equipment, grasping of cleated shoes for weightless astronauts, and a larger-appearing interior.

On May 14, 1973, Skylab 1 was launched on the final Saturn V to fly into space. The rocket left the pad in a rush of flame and noise and thundered its way east over the Atlantic Ocean, ultimately depositing Skylab into a 270-mile-high orbit. The plan was for controllers to activate Skylab's systems, deploy the solar panels, and perform checkouts. Then, once everything was in good order, the first crew to visit the station would launch atop a second, smaller Saturn IB rocket the following day. They would catch up with Skylab, dock, and enter the space station to begin their twenty-eight-day mission. That is what was *supposed* to happen.

But within an hour, ground controllers realized that Skylab was in trouble, and the launch of the crew was delayed. Houston was soon receiving puzzling telemetry from the space station. At first they thought it was a bad sensor or inaccurate readings on the ground, and did checks and double checks on their equipment. But that was not the issue. Skylab was in serious distress. The power from the solar panels was only about 20 percent of what it was supposed to be, and that was not enough to operate the space station. Worse still, temperatures were soaring inside, rapidly heating the interior to 130°F.[7] The combination of these two problems would wreck Skylab in short order if not addressed.

But before they could fix Skylab, they needed to know what had gone wrong. With a lot of detective work and intuiting, they diagnosed the major failures: the station had two folding solar panels that sat flush along its sides during launch. Once in orbit, the shields were supposed to swing away from the station and lock perpendicular to the hull. That had been the plan.

However, during launch, about fifty seconds into the flight, supersonic air had found its way under the leading edge of Skylab's heatshield and ripped it free. The port solar panel deployed prematurely, and aerodynamic forces ripped it off as well, the fragments fluttering into the Atlantic below. On the other side of the station, a strip of metal twisted and caught the remaining solar panel, preventing it from deploying properly.

Once Skylab was in orbit, four smaller solar panels unfurled from the forward end, a structure called the Apollo Telescope Mount. But when the controllers tried to unfold the two main solar panels, which provided the vast majority of the power, the starboard side was still stuck fast and the other side was simply *gone*. There was barely enough power to run the most basic functions.

As bad as this was, the missing heatshield was a worse problem. Its primary purpose after launch was to shade the station's hull from the blistering effects of direct exposure to sunlight in space. Most spacecraft in the 1960s were slowly spun in space to equalize temperatures across their hulls—the astronauts called it "barbecue mode." But Skylab was a research platform, and this was not an option—it was designed to fly in stable orientations—so a heatshield had been installed. Without it, the hull blistered and temperatures inside the station soared. It would eventually reach the point where the interior would be destroyed and the electronics would be compromised.

Pete Conrad was the commander of the first crew headed to Skylab.

Conrad was the veteran of the bunch, having flown on Gemini 5 and commanded Gemini 11 and Apollo 12. They had been scheduled to depart the day after Skylab launched, but this would now be delayed until NASA could figure out how to fix the problems, if they *could* be fixed. Failure to do this would result in the cancellation of the mission, something Conrad was not about to let happen.

Pete Conrad was a fixture in the astronaut corps. Most astronauts were test pilots—tough, taciturn, and cocky. But Conrad broke the mold. He was tough and cocky all right, but at 5'6" he looked more like a bantam rooster than the stereotypical rocket jock. He had earned his pilot's wings in the US Navy, and was a balding, gap-toothed extrovert, with a decidedly at-your-expense sense of humor. The other guys loved him, and he was one of the most respected members of the astronaut corps.

Conrad's crewmates were both rookies. Paul Weitz, another navy pilot, would remain in NASA to fly the space shuttle after Skylab. Joe Kerwin rounded out the all-navy crew and had an additional distinction to his résumé: he was a medical doctor, a handy person to have around during a long-duration space mission.

Conrad had driven the crew to perfection in training and was not about to let something as small as a launch malfunction stop him from completing the mission. He joined the huddle at Mission Control to see what plans were being made to get the station fixed and his crew on its way.

Mission engineers got together with the contractor who had built Skylab, McDonnell Douglas, along with the subcontractors, in an effort to find a solution quickly. They would have to work fast—at the rate that Skylab was roasting up there, they did not have a lot of time. Within days a number of possible fixes emerged, each with its own set of risks. None were foolproof. It was a matter of selecting the least complex and dangerous option that was likely to work.

For the overheating issue, they needed to protect the station from direct sunlight via some kind of curtain or shade. The second problem was the stuck solar panel. With the port panel gone, it was critical to fully extend the remaining panel, which was stuck partially open. They would need to design a fix for this problem as well, without even knowing exactly what had gone wrong. Whatever solutions they ultimately chose needed to be a) easy to operate and deploy, b) lightweight, c) robust enough to last for at least ninety days, preferably a year, d) be safe to fly with, and e) able to fit in the already cramped confines of the Apollo capsule while leaving the three crew members room to crawl over the mess and fly into space. They also had to be designed, reviewed, and built in a matter of days. It was the ultimate in careful improvisation.

Inside Skylab, food was spoiling and interior plastics were deforming. Wire insulation was beginning to melt. And possibly most dangerous of all were the noxious chemical vapors that could form as the plastics inside Skylab were heated. These potentially toxic gasses could scuttle the mission.

Many proposals came forward from contractors and NASA personnel. Cloth shades, inflatable structures, even spray paint and commercial wallpaper were suggested as ways to protect the slowly blistering hull.

Finally, the engineers at the Johnson Space Center came up with a design for partial sunshade—a large nylon parasol that could be shoved up through a small airlock built into the hull that could then be deployed much like a beach umbrella. A technician was sent into town to buy some fishing poles and the strongest nylon line he could find. The fabric came from a parachute found in a warehouse at JSC. The engineers rigged up the contraption and tested it inside an empty workshop, then hastily arranged a demonstration for the top NASA managers. The assembled bigwigs looked skeptical, but agreed—there were not a lot of other options that could be broken down and jammed under the seats in the command module.

The Marshall Space Flight Center in Huntsville, Alabama, also came up with a fix—an extendable nylon curtain that would unfold like a window shade. Two collapsible fifty-foot fiberglass poles would hold one end and push out the other. This device would cover a larger area, but had to be deployed via a spacewalk by an astronaut in a pressure suit. NASA, at this point, still did not have a lot of experience in EVAs in which the astronauts had to accomplish complex tasks. However, it was the only method that the engineers could think of to cover the remainder of Skylab's hull, which would still be exposed after the parasol was deployed.

And then there was the partially stuck solar panel. When the heat-shield and first solar panel were torn off during ascent, thin scraps of the wrecked metal shield became wrapped around the remaining solar panel. It was like aluminum roofing material—thin, sharp, and clingy. Worse yet, rivets had been used in its construction, and some of these had embedded themselves into the remaining solar panel. The panel was stuck fast, generating only partial power, and would need to be freed.

Some of the engineers went to a local hardware store and bought a tree-pruning pole and clipper head. A branch snipper costing perhaps $25 could be used to cut the metal restraining the solar panel, and a coil of plain rope could be used to pull it into place.[8] It was assuredly the cheapest critical component in the entire space program—this was a damn-the-torpedoes (and the endless studies), full-speed-ahead situation.

These Rube Goldberg contraptions were tested by fellow astronaut Rusty Schweickart in NASA's neutral buoyancy tank, a large pool where they practiced EVAs in a simulation of weightlessness.[9] When he had the procedures down, he showed them to Conrad and the crew, who thought it was hysterical. It looked like something Wile E. Coyote would fabricate to trap the Roadrunner, but they agreed to give it a shot. Kerwin joined Schweickart in the water and practiced for hours, learning as much as was possible in a simulated environment over just a couple of days.

All the while, Conrad fumed. He knew that people were doing their best, but *his* mission, *his* space station, was at risk. In one discussion he let his typical cool slip for a moment and spat, "Just get me *up there*, goddamn it!"[10] He knew if a way could be found to fix the thing, his crew could do it. But he needed to *get there*.

On May 25, eleven frantic days after Skylab had flown into trouble, Conrad, Weitz, and Kerwin climbed into their Apollo command module.

Due to the Saturn IB rocket being much shorter than the Saturn V, for which the new launch gantries had been built, it was propped up on a lifting platform they called the "milk stool."[11] It was an odd-looking arrangement—another seemingly ad hoc part of the Skylab mission—but it worked.

Conrad was gleeful as they left Earth, exclaiming, "Tally-ho the Skylab!" as they closed on the crippled space station.[12] As they got close enough for visual inspection he was more subdued. The station was a real mess, and his optimism ebbed . . . but they would try.

Conrad maneuvered the command module parallel to Skylab as if he were parking a sports car. Within the hour he got the okay to depressurize and open the hatch. Weitz assembled the branch cutters in the cramped confines of the capsule, and with Kerwin holding him by the ankles, Weitz stood out of the hatch and floated halfway out of the capsule in an attempt to wrest the pinned solar panel free. With the bulky pressurized space suit limiting his movements, he struggled to get the pole in place and grab the panel. He could see how the heatshield had gotten twisted around a section of the remaining solar panel, holding it fast, its dangling rivets snagging the panel. The harder he pulled and tugged on the pole, however, the closer the command module came to smashing into the station. Conrad had to fire the thrusters constantly to avoid a collision. This used much of their maneuvering fuel and left Conrad feeling uneasy. An impact would do neither spacecraft much good, and could end the mission then and there.[13]

Despite much tugging and pulling, the angles were just wrong and it wasn't working. After a twenty-two-hour day they prepared to dock and enter Skylab. Conrad maneuvered the capsule slowly toward the docking port, and the pointed probe on the command module slid into the collar on Skylab . . . and then, grinding in protest, slid right back out again. He gave it another try, and another, but the docking latches, designed to hold the spacecraft together, were not snapping into place. It was a day of frustration and failures.

Conrad backed away, concerned that if he rammed the station much harder he would ruin the docking port, thus ending Skylab's career before it started. Finally, the crew members put their pressure suits back on and depressurized the capsule again. In a daring and dangerous maneuver, Conrad nudged the capsule against the docking collar one more time as Kerwin crawled into the tunnel, removed the front hatch, and snipped a couple of wires in the docking rig—a decidedly unorthodox and ad hoc procedure. The rows of men at Mission Control were glued to their readouts.

The snipping of the wires up front disabled a safety protocol, and finally, with a series of bangs, the docking latches slammed shut, locking the craft firmly together. The trio hoped that the latches would let go when the time came to go home.

Still in their pressure suits, they sampled the air inside Skylab for toxins with hastily contrived test kits. Thankfully the results were negative—no poisonous gasses resided within. They were prepared to enter with their visors up, but floated into a wall of 130-degree stale air. Always practical, the crew stripped off their space suits and entered America's first space station in their skivvies.[14]

Once inside they moved quickly to deploy the jury-rigged parasol through an airlock—to the surprise of some of the doubters on the ground, it worked, and temperatures started to drop inside. Then being careful to observe extreme limits on power, they began to configure the station, though food and water were still consumed cold and very few lights and circuits were turned on due to the massive power deficit. That would have to be dealt with, and soon.

After two weeks of these deprivations, they prepared for an EVA to fix the solar panel. NASA had encountered significant challenges with orbital EVAs, and there were real risks associated with the repair procedure. The most recent experience with real work and tools in a zero-g environment had been during the Gemini program years before. Only on the final flight of that program had Buzz Aldrin demonstrated these abilities successfully. During Apollo, in-space EVAs had involved only very simple tasks, and the plan for Skylab called for only slightly more ambitious work. Now NASA and Conrad's team would play for much higher stakes. The future of their mission, and the two subsequent crews scheduled to inhabit Skylab, depended on their success.

Houston sent up revised and updated instructions for the EVA procedures to Skylab's printer station. The printout was almost fifteen feet long. How much attention Conrad paid to it is not recorded, nor are his words when seeing it, but they were likely to be salty.

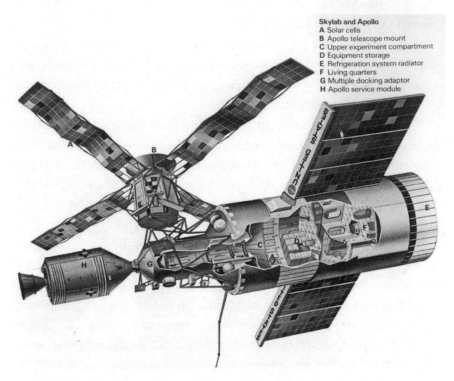

Skylab and Apollo
A Solar cells
B Apollo telescope mount
C Upper experiment compartment
D Equipment storage
E Refrigeration system radiator
F Living quarters
G Multiple docking adaptor
H Apollo service module

Fig. 18.1. An early illustration of Skylab as it should have appeared had every-thing gone as planned. As it turned out, the solar panel at the top of the illustration was torn away during launch, leaving the space station with grossly reduced power. Image from NASA.

Conrad and Kerwin exited Skylab the next day to begin work on the snagged solar panel while Weitz remained behind where he could observe their progress. Each of the two EVA astronauts struggled with his equipment—fiberglass poles, rope, the branch cutter, and miscellaneous odds and ends. The airlock was not really designed for space-suited astronauts armed with municipal gardening equipment, but they made it work.

Kerwin extended the fiberglass pole to its twenty-five-foot limit and mounted the cutter on its tip. Since the outside of Skylab was smooth and devoid of handholds or grips (they had never planned to work on this side of the station), they would hook the cutter onto the offending metal shroud with the long pole and close the jaws halfway, enough to grip but not enough to sever. Then they could secure the other end of the pole to the station and use it as an ad hoc handrail. Assuming this worked, which was by no means certain, Conrad would then use the contrived handrail to pull himself to the solar panel, tie off a piece of rope, come back, and tie off the other end of the rope to some part of the station as yet to be determined. Then he and Kerwin would produce enough tug on the rope—somehow—to finish cutting through the mangled heatshield, which would then spring free. The stuck solar panel would swing into its open position and lock. That was what they had practiced.

Down in Mission Control, the soundtrack of two grunting, wheezing, and wise-cracking astronauts did not lessen the worry on the ground. Planners had discussed the procedure at length, and many had been concerned about the "what ifs." What if the metal restraining the solar panel could not be cut? What if it tore free and flew over and cut open one of the astronaut's suits? What if a rope snapped and whiplashed, somehow tangling with an astronaut, dooming him? What if something else broke free and somehow injured one of the spacewalkers, disabling him and forcing his companion (presumably also injured) to leave him outside? What if, what if?

In space, Conrad did not appear to share their concerns. Nothing was going to prevent the success of the mission short of an immediate danger to himself or a crewmate.

Kerwin set about trying to slip the cutting jaws over the heat-shield fragment, but could not steady the twenty-five feet of extended, whipping pole enough to grab hold. It was frustrating and risky. Then Conrad noticed a D-ring attachment point protruding from the side of Skylab—there had not been one on the simulator on the ground, but here it was, and that was all that mattered. They tied one end of the errant rig to the hull, steadying it, and soon had the jaws of the cutter firmly clenched around the metal shield.

While Kerwin caught his breath, Conrad worked his way along the pole, out to the end of the solar panel, trailing a length of rope. He tied it off and pulled himself back over to Kerwin, only to have the cutter jaws pick that moment to jerk and cut the restraining metal free. The panel swung out about twenty degrees, and the pole drifted away—as did Conrad, "ass over teakettle," as Kerwin later described it.[15] Conrad reeled himself back with his tether, and the two men assessed the repair. The heatshield was clear, but now a damping strut was preventing the panel from opening out to the full ninety degrees.

The strut would have to go, but that was not a part of the procedures worked up by Mission Control. The astronauts quickly devised a new course of action. With the rope tied off at both ends—one on the far side of the solar panel and the other on the D-ring—Conrad and Kerwin climbed underneath it and on Conrad's cue both stood quickly upright, putting the rope under stress. There was a groan as the metal torqued, and then the damping strut let go, the solar panel swung free, and once again the astronauts cartwheeled into space.

Laughing and dangling from their tethers, they hauled themselves back to Skylab for the last time. Their work was done. Only after they completed the task did they fully explain the unorthodox procedure to the nervous controllers on the ground—not the least of which were the doctors, who seemed to be convinced that anytime a man's pulse rate topped 140, he might explode. Conrad and Kerwin returned to the airlock, went inside, stripped down, and treated Houston to another stream of colorful language as Conrad narrated the experience to Weitz, laughing all the while. It was like listening to a late-night blue-humor comedy act in Vegas.

Temperatures stabilized, the power flow was raised enough to be workable, and Skylab was now sufficiently stabilized to enable the remainder of their twenty-eight-day mission. On-the-spot ingenuity and NASA know-how had rescued the crippled space station. The crew was now able to fully concentrate on science experiments, exercise regimens for zero-g conditioning, and other tasks that they had fore-stalled for two weeks.

For Conrad's efforts he received the Congressional Space Medal of Honor. In subsequent years, despite being the third man to walk on the moon, he would always count Skylab as his favorite mission. He probably laughed more, and swore more, on that mission than any other time in space.

Two more crews occupied Skylab after Conrad and his shipmates. The second mission was commanded by Conrad's lunar module pilot from Apollo 12, Al Bean, who stayed on the station for fifty-nine days. The third and last crew to inhabit Skylab stayed eighty-four days, a record that no American would break for twenty-five years.[16]

Skylab crews conducted over 2,000 hours of scientific and medical experiments, and took almost a quarter million pictures of the sun and Earth. It was the most advanced solar observatory of its time, allowing for observation high above the Earth's atmosphere. It was also the first platform that allowed extended biomedical observation of astronauts in zero-g conditions.

After the third and final crew left in February 1974, Skylab was mothballed. There were plans drawn up for future missions, but these would depend on the space shuttle for transportation, since no more Saturn rockets were allotted to service Skylab. The station was left with three months of water and even more oxygen for a crew of three, and some supplies could have been replenished. But the space shuttle continued its long slide into a delayed flight schedule, finally making its first ascent in 1981. By then it was too late for Skylab.

The station had been left in a high orbit, at about 270 miles. It was expected to be stable there until the early 1980s, and by then NASA would have decided what to do with it. But a higher-than-normal cycle of solar activity heated the Earth's upper atmosphere more than expected, increasing drag on the station even at that altitude. Without an available spacecraft to fly up and boost it to a higher orbit, Skylab was doomed.

On July 11, 1979, Skylab came home. NASA had managed to maintain contact and had some control over the station's orientation, and did its best to initiate reentry such that any debris would fall over unoccupied ocean areas. However, the reentry profile was not quite what had been expected, and Skylab came down over Australia, fortunately almost as unpopulated as the oceans for much of its interior. Nobody was hurt by the falling debris.

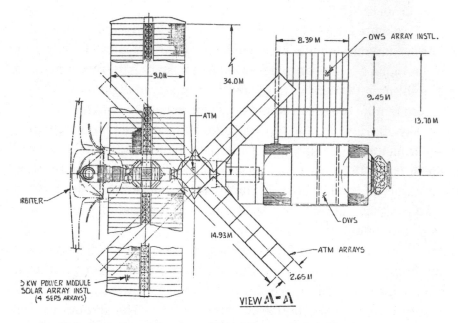

Fig. 18.2. Illustration of the planned Skylab "rescue" mission by the space shuttle. The shuttle (to left) would have docked with Skylab, as indicated to the left, and carefully boosted the space station to a higher orbit. As it turned out, the shuttle fell behind schedule, and Skylab's orbit decayed far more quickly than anticipated. It reentered and broke up in 1979. Image from NASA.

Skylab was, in the end, a spectacular mission, created primarily from repurposed existing infrastructure and technology. Not until the Soviet MIR space station flew in 1986 did anything even come close to Skylab in terms of research and crew longevity. But perhaps the most important legacy was the inventiveness, courage, and raw tenacity of its crews, and the many controllers, engineers, and technicians who supported them. Their efforts set a high bar for on-site innovation and command decisions in space.

CHAPTER 19

NEAR MISSES: DANGER STALKS THE SPACE SHUTTLE

UNCLASSIFIED

As America's longest-flying spacecraft—by a wide margin—the space shuttle faced numerous dangers during its thirty-five years of spaceflight. We all know of the two shuttles lost in flight, when *Challenger* disintegrated during ascent in 1986 and *Columbia* broke up upon reentry in 2003, but we're less aware of the other near misses that occurred during its long and fruitful lifespan.

To set the stage: the shuttle we saw flying for over three decades was not the shuttle we were supposed to have. Original designs for spaceplanes date back to Sanger's *Silverbird*, and in fact, the US shuttle was originally a similar, though larger, design. As envisioned in the 1960s, it would have had stubby wings like *Silverbird* and the X-15, and a wide tail with winglets on the tips. But many factors conspired to change that design to the shuttle that ultimately flew.

Early designs for the space shuttle called for a completely reusable system. The goal was to create routine and affordable access to space, saving large sums over the cost of expendable rockets and spacecraft like the Apollo Saturn V and command/service module system. The US Air Force's X-20 Dyna-Soar was not dissimilar, though in that system, only the orbiter would return for reuse. Dyna-Soar research and experience with the X-15 were both beneficial to the shuttle program.

Dyna-Soar was shut down in 1963, though the X-15 flew until 1968. By that time, early drawings and descriptions of a true space shuttle were being circulated between the aerospace contractors and the government. There were many variations, but most involved a large, winged, reusable booster/plane that looked like a giant shuttle itself, flown by one or two pilots. This would launch with the orbiter (the actual term for the shuttle we know) attached to its back. At altitude, the booster would release the shuttle, which would continue into orbit. The booster would then fly back to land on a runway, to be refurbished and launched again. The orbiter would complete its mission, reenter, and land on a similar runway to be treated in a similar way.

Engineers were, at the time, incredibly optimistic about turnaround times, and it was thought that this system could be cleaned, inspected, fueled, and flown again within days or weeks at the most. This turned out to be fiction—the space shuttle was a vastly intricate and complex machine that required expensive and time-consuming maintenance and repair between each flight, with spare parts and even engines swapped out between the four orbiters with regularity.

The shuttle designs of this era were smaller than the modern space shuttle—somewhere between Dyna-Soar and the shuttle the US ultimately flew—capable of hauling only about 12,000–20,000 pounds of cargo into orbit. Two people within NASA were largely responsible for competing designs: George Mueller, who led much of the Apollo program, and Maxime Faget, the designer of every spacecraft NASA flew from Mercury through the eventual shuttle.

In 1969, President Nixon asked Vice President Spiro Agnew to form a new Space Task Group (STG), which would determine the post-Apollo goals for NASA. Of course, NASA had many ideas of its own that had been studied extensively, mostly Apollo follow-ons that capitalized on the enormous engineering work that had been put into creating that spaceflight system. A permanent moon base and expeditions to Mars were on the table. But the STG knew that NASA's budget was being cut and such large-scale efforts would not be tenable; a reusable spacecraft that could fly to orbit and operate there for extended periods, while delivering various cargoes, and fly back to Earth for reuse was eventually chosen. But after extensive study, even this system, as envisioned, would have been beyond the spending limits now shackling NASA.

As the studies continued, and the gradual awareness crept in that even a space shuttle program was at risk, the air force was invited to join the effort. With the death of the X-20 Dyna-Soar in 1963 and the cancellation of the Manned Orbiting Laboratory in 1969, there was still space money in the air force's budget without an approved program to spend it on. But the air force had different needs and expectations than NASA, and it was an uneasy relationship.

For military purposes, the air force needed a shuttle that could carry more cargo. The final design would be capable of lifting almost 60,000 pounds to low Earth orbit and 28,000 pounds to a polar (north–south) orbit, one generally preferred for military purposes. The orbiter would also need enhanced cross-range capabilities, the ability to fly well off the equatorial track that was the previously accepted norm for the NASA designs. The ability to launch and land at Vandenberg Air Force Base on the California coast was one requirement, as were the plans for deployment of various military hardware, reconnaissance flights, and more. The shuttle grew heavier, had larger wings, a taller rudder, and became more expensive. The air force would have to step up and make good on its financial promises for the program to work.

The final design abandoned the reusable, piloted booster for reasons of cost and complexity. In its place were two huge solid rocket boosters (SRBs), which provided the majority of the power to get the orbiter into space, and a large external tank, feeding three liquid rocket engines inside the orbiter. The SRBs fell away after about two minutes of flight, to parachute into the Atlantic Ocean for recovery and reuse (they ended up being expensive to refurbish, nearly negating their value as "reusable" components). The external tank (ET) was a one-use part of the system, and was cut loose to disintegrate during reentry after the orbiter had enough velocity to reach orbit. Then the orbiter, at the completion of its mission, would glide to a landing at the Kennedy Space Center, Vandenberg Air Force Base, or, in case of complications or bad weather, Edwards Air Force Base, also in California. The orbiter did bring the three expensive rocket engines, the Space Shuttle Main Engines (SSMEs), back with it, which represented a savings.[1]

The overall development budget was about $5.6 billion.[2] Budget overruns due to delays and engineering difficulties pushed this to about $6.7 billion, a small overrun when developing a new spacecraft system, and still far less expensive than Apollo. Congress complained endlessly, but it was, in the final analysis, a reasonably cost-effective system to build. Operating it was another thing altogether.

The air force justified its involvement by, somewhat reluctantly, agreeing to fly all its orbital payloads on the shuttle. The long list of expendable rockets then in use—the Atlas, Titan, and Delta rockets—would no longer be used for air force payloads. Various "upper-stage" rockets would be required to finish the journey for many military pay-loads, and these would ride in the shuttle's payload bay along with whatever the air force was sending into space. One of these power plants, a cryogenically fueled rocket stage called the Centaur, was redesigned to be carried in the shuttle instead of acting as the upper stage of conventional rockets. This could have been effective, but was canceled after the loss of *Challenger* due to the perceived dangers of carrying explosive rocket fuel inside the payload bay.

The shuttle orbiter contract was awarded to North American Rockwell (formerly North American Aviation) in California, and the solid rocket boosters to Morton Thiokol in Utah. The external tank was built by Martin Marietta at NASA's Michoud assembly plant in Loui-siana (former home of the Saturn V's first stage). The goal was to have the entire system flying by 1978. But after the usual teething troubles and the effects of a decade of ever-slimming NASA budgets, the first flight was not until 1981.

From the beginning, the shuttle was in many ways a disappointment. The cost of accessing space was not reduced by the system, and the shuttle actually ended up being more expensive than most of the rockets it was supposed to replace. Early on, the cost of one pound delivered to orbit by the shuttle was suggested to be in the hundreds of dollars; in the end it was still about $8,000. And everything sent into space had to be engineered to be safe for a crewed spacecraft delivery; formerly most of these payloads only had to meet the standards for unmanned rockets. The shuttle did not turn out to be a bargain for its customers.

The rate of launches also turned out to be a letdown. Flight rates were at one time projected to be about one per week, or perhaps even fifty-four per year. In reality, they never did much better than one per month, and the average over the life of the program was about one every three months.[3] The orbiter had a lot of areas that needed attention after each flight to make it safe for reuse, and even with a fleet of four, the downtime between uses was significant. Every component needed to be checked and verified for flight status. The heat-resistant tiles were a problem as well, though this issue lessened in severity as the program matured. In early tests and flights, the orbiters came back missing tiles, and others were damaged and also needed to be replaced. Each tile averaged a few inches on each side, and had to be custom made specific to its placement on the orbiter. They worked, but were incredibly fragile, and making them adhere to the orbiter proved far harder than envisioned.

The SSMEs, the main engines on the orbiter, also proved to be more of a challenge than originally thought. These were high-performance machines that burned high-energy fuels—liquid hydrogen and liquid oxygen. The turbines that moved fuel through the power plants spun at incredibly high rates, and were high-maintenance items—with impeller blades spinning upward of 28,000 rpm for the liquid oxygen and 35,000 rpm for the liquid hydrogen, any faults in these components (a loose blade or piece of detritus inside the turbine) could spell disaster in flight.

As mentioned, the SRBs had their issues as well. They were complex machines despite their essential simplicity as giant sky-rockets, and refurbishment after being retrieved from salt water was more time consuming than envisioned. In the end, the shuttle program cost close to $200 billion, or about $1.4 billion per flight (spread over 135 flights).[4] It was expensive.

But NASA had an agenda, and it was heavily skewed toward keeping the flight schedule up. For a few years, once the shuttle had been declared "operational" after only a few flights, the space agency drove itself to deliver on its promises. Starting in April 1981, *Columbia* was flown twice in that year, then three more times in 1982. *Challenger* flew for the first time almost exactly two years into the program in April 1983. By 1985, all four orbiters were in use—*Columbia*, *Challenger*, *Discovery*, and *Atlantis*. The program reached its peak in that year, nine launches in twelve months. This was not even a shadow of the planned fifty-four launches per year, but was as good as it got. And, rather than streamlining procedures and amortizing costs, the frantic pace of the program resulted in shortcuts, relaxation of flight rules, and rapid recycling of parts between the orbiters. And all that added up, in part, to *Challenger*'s demise.

The details of the *Challenger* accident are well documented. The short version is that on January 28, 1986, the shuttle was launched in the coldest weather ever attempted. One flight rule after another was waived by NASA management, and despite the cold snap and many concerned engineers and technicians, at 11:39 a.m. EST, the spacecraft launched into a clear sky. Seventy-three seconds later, it disintegrated just eighteen miles up. Emergency procedures were implemented, but the orbiter was a total loss, plunging at terminal velocities into the Atlantic in pieces. The SRBs continued to burn and ascend until destroyed by the air force range safety officer. The crew cabin of the orbiter smashed into the ocean relatively intact, but was crushed by the force of impact. All seven astronauts died.

During the resulting shutdown of the program, investigators determined that due to a combination of poor design and the cold weather, a rubber O-ring that sealed a joint between sections of the left SRB deteriorated due to burn-through. The hot gasses of the burning solid fuel spewed out of the joint like a blowtorch, severing the lower strut that affixed the SRB to the external tank. When that strut let go, the SRB pivoted on its remaining upper strut, and its tip punctured the external tank. With fuel raining down from the top of the ET, and with structural loads shifting and the aerodynamics of the vehicle compromised, the shuttle broke up and the fuel ignited, creating a huge fireball. While the crew cabin of the orbiter was blown clear, and continued ascending until it arced into the ocean, survival was impossible. There was no abort system for that phase of the flight—when the SRBs were burning, you were along for the ride no matter what.

As the famed physicist Richard Feynman, who was a member of the investigation board, pointed out, the O-rings had not been tested below 50°F. They became brittle when cold, as he famously demonstrated by immersing a piece of one in a glass of ice water and snapping it in two. He ended his summary with, "For a successful technology, reality must take precedence over public relations, for nature cannot be fooled."[5] There was no better summary of the relaxed safety culture that had taken hold at NASA prior to the accident.

The other major accident of the program was the disintegration of *Columbia* during reentry on February 1, 2003. On that flight, the shuttle suffered a burn-through of the airframe on the leading edge of the left wing. The wing structure melted, compromising the aerodynamics of the orbiter while flying at Mach 18, and it broke apart some 200,000 feet in altitude. Again, there were no survivors—seven more astronauts died.

During the launch of *Columbia* on January 16, eighty-two seconds after liftoff, a piece of insulating foam the size of a small suitcase broke free from a support strut on the external tank, striking the leading edge of the left wing. That part of the wing is covered with reinforced carbon composite, or RCC, a highly temperature resistant material that is strong but not flexible. Subsequent investigations indicate that this foam strike punched a hole in the RCC about six to ten inches wide. This hole is where the hot reentry gasses bled into the wing, destroying the orbiter.

The foam strike was spotted on video shortly after launch. Some of the engineers consulted felt that there was a chance of wing damage that could be potentially catastrophic. But NASA management limited the investigations of any potential damage during the flight, reasoning that nothing could be done to fix it in any case. *Columbia* orbited for nearly sixteen days with a gaping hole in its left wing that would doom the crew upon reentry.

Once again, shuttle flights were halted while the accident was investigated, this time for two years. Changes were made to the program to enhance safety: shuttles would now use the robotic arm to survey the spacecraft once in orbit, seeking any damage to heat-resistant materials. Additionally, a second shuttle was kept on launch standby unless the mission profile allowed the shuttle to return to the International Space Station in case damage was found.

These are the two major incidents that caused the loss of two shuttles, fourteen astronauts, and years of flight time. Each resulted in wide-ranging investigations of NASA and the shuttle program, and major changes to safety protocols and procedures. These are the accidents we have seen endlessly portrayed in the news, books, articles, and documentaries. But other, smaller "anomalies" occurred during the thirty-five years of shuttle operations that warrant a closer look. None of these precipitated disaster or loss of life, but they do indicate the many dangers of spaceflight and, in particular, some of the shortcomings of the space shuttle's design.

In July 2005, over two years after the loss of *Columbia*, *Discovery* was moved to the launchpad. The accident had been scrutinized and studied to every last detail. Every variable had been considered. Chunks of foam had been shot at the leading edge of wing structures. The insulating foam had been frozen, heated, and blasted with hot slipstream simulations in attempts to keep it intact during launch. Countless computer analyses and simulations had been run. People had given up any semblance of normal lives to identify the issues and stop problems before they became problems. Everyone was confident that they had it licked—the fault had apparently been the result of problems with the installation of the foam. There were small voids between the insulating foam and the external tank and support struts that allowed the foam to peel off. New procedures were introduced, and the problem was apparently solved.

When they launched *Discovery* that morning, everything appeared nominal. It was a picture-perfect liftoff . . . but within hours that picture turned dark. A technician scrutinizing still frames of the launch video noticed something. A chunk of insulating foam had struck the leading edge of the left wing . . . again. It was not clear if there was damage or not, so the crew was told to investigate the wing using the camera on the robotic arm. Nothing showed up as damaged or even scuffed, and the reentry proceeded safely and on schedule.

So what went wrong? After extensive reanalysis, the engineers realized that it was not poor installation processed that was causing these accidents, but "thermal cycling." The very act of filling the external tank with the icy cold cryogenic fuels could cause the foam's bond to weaken and the insulation to separate from the external tank. Repeated fill/drain cycles of the cold fuels, which happened when there were launch delays, just made the problem more likely to occur. Procedures were reviewed and improved again, with any visible cracks or swellings of foam insulation x-rayed and microscopically examined from then on before flight. And the shuttle grew ever more expensive to fly.

These two events were not the first time that the shuttles had experienced issues with falling insulation damaging the orbiter. There were a number of others. In 1988, a foam strike from SRB insulation damaged over 700 tiles on the underbelly of *Atlantis*. Some of this was visible to a video feed from the robotic arm, but there was nothing to be done. This was also a military mission, so communications were restricted and the video feed to Mission Control encrypted and in low resolution for security reasons. Apparently it did not look excessively dangerous to the engineers in Houston (and in 1988, there was nothing to be done but come home anyway), but it surely looked bad to the crew. "I will never forget, we hung the (robot) arm over the right wing, we panned it to the (damage) location and took a look and I said to myself, 'we are going to die,'" recalled shuttle commander Robert "Hoot" Gibson. "There was so much damage. I looked at that stuff and I said, 'oh, holy smokes, this looks horrible, this looks awful.'"[6]

After *Atlantis* landed and the crew got a look at the side and underbelly of the orbiter, the commander's reaction was validated. A large swath of the orbiter's insulating tiles were chewed up, gouged, torn, and in some cases just missing. One patch of aluminum hull near the nose was partially melted. All that from one chunk of insulation sliding along the orbiter's underside at launch.

In fact, this kind of incident occurred about eighty times throughout the shuttle program. Proper investigation of the warnings could have prevented the *Columbia* accident, but they were instead noted and largely put aside.

There had been multiple warnings prior to the *Challenger* accident as well. In January 1985, *Discovery* flew a mission called 51-C. When the SRBs were recovered and disassembled—they came apart into four separate tube segments—severe blowby was seen on some of the O-rings on both SRBs. The rubber gasket, the culprit in the *Challenger* accident, was severely charred and burned away, allowing hot gas to reach a secondary seal (this too failed on *Challenger*). The *Discovery* launch was also the coldest to date, at 53°F.[7] This alone should have rung alarm bells when *Challenger* was attempting to launch on a morning that was only 28°F.

After the *Challenger* incident, the joints between the SRB segments were redesigned to add another O-ring, and also had a rim-capture feature—basically, rather than allowing the joint to bend open if the booster flexed, it would tighten. This is how it should have been designed in the first place.

And there were dangers while in orbit as well. Low Earth orbit, where the shuttle operated, is filled with space junk. Debris from exploded satellites, spent rocket boosters, meteorites, and dozens of other sources, orbit Earth at the same altitudes as the shuttle did and the International Space Station does today. And the numbers are frightening: over 500,000 pieces of space debris are tracked by the air force, and it's estimated that over 170 million bits smaller than a half inch are roaming Earth orbit.[8] And it doesn't take much to damage a spacecraft--a piece of detritus the size of a BB carries the destructive punch of a bowling ball traveling at 60 mph. Even a fleck of paint can be dangerous—in 1983, *Challenger* had a major chunk of glass knocked out of one of its front windows—it looked like it had been hit by a small bullet—that turned out to be a dime-sized paint chip. At orbital velocities, even such tiny, low-mass objects are to be feared. Fortunately, shuttle and space station windows have multiple layers of tough fused silica glass to protect them.

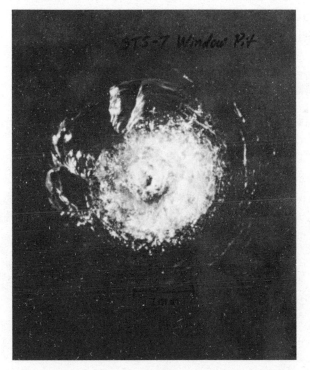

Fig. 19.1. The shuttle's front windows were damaged numerous times throughout the program. This image shows the result of a tiny piece of orbiting debris—probably not much larger than a BB—impacting the window. While such damage never went past the front glass pane (the windows were layered), it was a jarring reminder of the dangers of spaceflight. Image from NASA.

But these small bits or debris have taken a toll—NASA replaced windows dozens of times during the first seventy missions of the shuttle, at about $40,000 per replacement.[9]

And there are larger threats as well. In January 1999, the shuttle *Endeavor* had to make an emergency maneuver to avoid a collision with a 350-pound air force satellite that they had lost contact with a year and a half earlier. They missed it by about six miles, but that's too close for comfort in space. Ultimately, the shuttle program reoriented the orbiter while in space to minimize damage from such events, flying engines first (they were not critical at that point, having done their work during launch, compared to windows on the other end). This issue remains a problem for the International Space Station, which has to occasionally reposition itself to avoid being damaged by drifting space debris.

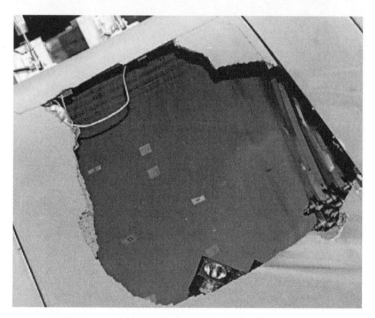

Fig. 19.2. After the loss of *Columbia* in 2003, tests were conducted to see what might have occurred to doom the flight. Chunks of insulating foam, identical to those that were falling off the external tank during launch, were shot at the leading edge of the wing by a cannon-like device. This is the shocking result. Thereafter, inspections of critical parts of the shuttle's heat shielding were conducted once the crew reached orbit. Image from NASA.

There were some main engine failures during the shuttle years, but none that resulted in a failed mission or death. In July 1985, during the launch of flight 51-F, one of the main engines shut down just forty-five seconds into flight due to an erroneous temperature reading. A second engine was about to shut down for the same reason, which could have been a disaster, or at least a very challenging abort. But a quick-thinking flight controller blocked the shutdown. Regardless, the flight performed what is called an abort to orbit, which generally means "keep flying until you reach orbit; then we can decide what to do."

In July 1999, a launch of *Columbia* nearly resulted in the shutting down of two of the three main engines. A short circuit knocked out two engine controllers, computer boards that run the rocket engines. They both immediately switched to backup systems, but had those backups failed, *Columbia* might have been forced to ditch into the ocean. This was not thought to be survivable.

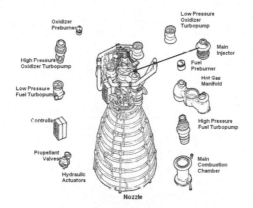

Fig. 19.3. Schematic of the Space Shuttle Main Engine or SSME. A complex and high-performance unit, the SSME has performed without a dangerous malfunction throughout 135 flights. Image from NASA.

The logbooks of the shuttle missions are full of near misses, but these are some of the notable ones. The space shuttle program provided service for thirty-five years and 135 flights, with a loss of fourteen astronauts and two shuttles. It is estimated that the same money could have paid for six launches of the Saturn V each year, with two of those flights being lunar missions. That said, the shuttle did ultimately build most of the International Space Station and provided transit to and from the complex for over a decade. It was the longest-serving spacecraft in US history, and was only outpaced by the Russian Soyuz capsule.

CHAPTER 20

SHOWDOWN IN SPACE: FIREARMS ON THE MOON

CLASSIFIED: *DECLASSIFIED IN 1971*

Where humans go, weapons will follow. This dictum had held true throughout recorded history, from the ancients through the colonial adventures of the West in the eighteenth, nineteenth, and twentieth centuries, and in national conflict up through today. But space has been remarkably weapons free for the most part, and remains so in the twenty-first century. There was experimentation with space-based lasers and kinetic anti-satellite weapons in the 1980s and 1990s, and the Chinese military launched an ASAT weapon test in 2007, successfully destroying an orbital target and adding countless bits of dangerous debris to the already crowded realm of low Earth orbit. But with regard to firearms—pistols, rifles, mortars, mines, grenades, and the like—the spacefaring nations have shown remarkable restraint. This is partially due to the space treaties of the 1960s, which ratified agreements that space was to remain a peaceful endeavor, and partially to the lack of perceived need—neither the US nor the Soviet Union foresaw scenarios likely to call for space-rated small arms. But that does not mean that smaller groups within the armed services would not prepare for the possibility.

There is one exception, a type of firearm that has been carried into space for many years. Because the Soviet (and now Russian) Soyuz capsules return to dry land, and because this can mean anything from landing in the targeted landing zone to the wilderness of the Urals or the wastes of Siberia, cosmonauts have long had sidearms packed in their survival kits. Even in the era of the International Space Station, where such weapons are forbidden within the station, a pistol resided inside the survival kit packed in the capsule for possible defensive use after landing through at least 2006.

The longest-serving specialty weapon was called the TP-82. Prior to the deployment of this gun, standard-issue military pistols were carried in Soviet spacecraft after the landing of Voskhod 2 in 1965, 600 miles off target in the Ural wilderness. On that mission, the two cosmonauts carried a small handgun, but this was not deemed appropriate for dealing with bears and other threats likely to be encountered in the Siberian wilderness, so in 1982 a new, purpose-designed weapon was ordered.

The TP-82 is a large pistol with three gun barrels: the top two are short, smoothbore shotgun barrels that fire slim 40-gauge shells, and the lower barrel fires an AK-47 round. The pistol can be used for defense and hunting, and the shotgun chambers can fire distress flares high into the sky to facilitate rescue. It also has a detachable shoulder stock for conversion into a light rifle. Official reports state that the weapon's use was discontinued in 2007, supposedly due to the specialized ammunition stores being long out of date. It was replaced with a more traditional sidearm. Then, after US astronauts began flying on Soyuz after the decommissioning of the shuttle in 2011, use tapered off. When pressed, cosmonauts and US astronauts who were trained on the weapon's use (even before the end of the shuttle era, some US astronauts and many international crew members flew on Soyuz) have remained tight-lipped about the firearm, since its very existence is a bit provocative.

It should be noted that this is not an assault-style, rapid-fire weapon—far from it. The triple barrel opens and tilts upward to allow each cartridge to be loaded one at a time, so its maximum fire ability would be three shots—two small shotgun shells and a single bullet—before it needed to be opened and reloaded. This is a nod to simplicity of design—it is a weapon that will never jam or misfire due to a poorly fed round, even if dropped in the mud or snow of an errant landing zone. Still, when you're discussing a weapon that is stored in the confines of a small capsule or a thin-walled space station, even a single-shot firearm can make people nervous.

A companion belt contained ammunition, flares, and a short machete (which also doubled as the rifle stock). All this was packed inside a small metal case carried between the seats of the Soyuz. The weapon had been used for training, and several were given to Soyuz mission commanders after their flights, but to date the weapon was never fired in anger.

These are the sole known deployments of sidearms in human spaceflight. But back in the 1950s, before the formation of NASA and when the military was in charge of trying to get Americans into space, it was a different world, and defense of the high frontier was prominent in a number of planners' minds. There were concepts for military space stations (see chapter 6) and army bases on the moon (see chapter 2), among others. Even the designs for orbital interceptors such as Blue Gemini and Dyna-Soar (chapters 8 and 12) discussed the possible use of weapons on these spacecraft, though apparently none made it past the proposal stage. But there was one group of hand-carried firearms that were proposed, and considered, as late as 1965, long after the Horizon moon base was just a yellowing-paperwork memory.

One study—more of a paper, really, at twenty-nine pages—came out of the headquarters of the US Army Weapons Command in Rock Island, Illinois, also the location of the Rock Island Arsenal, the largest military weapons manufacturer in the United States. The Rock Island Arsenal is the prime manufacturing site of armaments for the US Army, and has been since the 1880s. Weapons ranging from howitzers to heavy artillery to small arms, including pistols and rifles, have been manufactured there. So it is no surprise that enterprising and forward-looking minds were considering new types of firearms in the post-WWII world of technology and possible flights into space. Guns would have a place in this new frontier.

The study was rather whimsically named "The Meanderings of a Weapon Oriented Mind When Applied in a Vacuum Such as the Moon" (presumably the author meant to imply that he was applying his mind to weapons being fired *on* the moon, not that he was applying his mind within a lunar vacuum.) It was quickly classified, then approved for gradual release in 1972. It is labeled "Directorate of R&D, Future Weapons Office." Clearly they took their work seriously.[1]

It's difficult to fully convey the unusual spirit of the paper, so here is the preface in the author's words:

PREFACE

1) The purpose of this brochure is to stimulate the thinking of weapon people all the way from those who are responsible for the establishment of requirements, through those who are responsible for funding, to the weapon designer himself.

2) Although the primary purpose of man in space (on the moon or other planets) will not be to fight, he requires the capability to defend himself if necessary. There may be other countries desirous of preventing U.S. access to the moon and other planets. If space is truly for peace, we must be strong there just as we are on earth.

3) Because of the entirely new and different environment and conditions facing man in space, we cannot wait until the eleventh hour to "crash" a weapon program through with any hope of success, for we may even now be standing on the edge of the battleground of Armageddon. To quote our President before he ascended to the Presidency, "Space is infinite. Man's knowledge of space is finite. The sum of our understanding is not yet sufficient for us to comprehend how vast the dimensions of our ignorance."[2]

"If space is truly for peace, we must be strong there just as we are on earth." You simply have to love this quote. The US, Soviet Union, and many other nations were less than two years from signing the Outer Space Treaty of 1967 (it is still in effect), and there had been no sign of direct conflict in space since the beginning of the space age. But admittedly both the US and USSR were quietly pursuing various military space programs, some involving interception of orbiting enemy assets, so maybe the author was not as far off as he seems.

There follows a fairly detailed discussion of temperature extremes, the behavior of metal in a vacuum (it was feared that moving parts in a gun might weld themselves together), sight lines for projectiles in low-gravity environments, poor performance of lubricants in the cold lunar environment, and so forth. It is a piece of bizarre but responsible engineering, given what was known at the time. The author then summarizes his "meanderings" thus:

CONCLUSION

If the moon and other planets are explored and possibly colonized, the world could eventually see a second evolution of weaponry and protection therefrom. Visualize starting with a weapon capable of penetrating thin skinned vehicles. The vehicles then get thicker skin. The weapons then attain a greater penetrating capability. The vehicles get even thicker skinned until the weight and cost thereof becomes insurmountable. The weapons attain longer ranges, etc., etc., etc. This proceeds through the mortar, howitzer, gun and tank stages until eventually you have missiles, antimissiles and nuclear weapons much as the earth had prior to World War III.[3]

It is not clear if the author meant WWII and committed a typo, or considered the Cold War to be tantamount to WWIII, or was projecting a war in space as WWIII. In any case, he was discussing how to build guns to be shot on the moon, so the possibilities for imaginative improvisation are vast.

Of the weapons postulated for use in space, which in this paper specified a lunar environment, here are a few highlights:

SPIN STABILIZED MICRO GUN: A very Buck Rogers–looking pistol that used a traditional propellant, probably gunpowder, to propel a traditional-looking bullet. The ammunition is shaped like an ordinary .50-caliber bullet, but is a tiny .14-caliber round, slightly smaller than a BB. The gun holds fifty rounds and is semiautomatic in operation—one trigger pull equaled one round fired. The muzzle velocity would be 3,000–4,000 feet per second, or about three times the speed of a traditional .22 rifle round. The pistol was nearly two feet long and had cool-looking heat-shedding fins on the barrel, since when operating in a vacuum, heat would not be removed from the weapon as readily as it would be when operating on Earth in an atmosphere. There was a small fold-down grip on the front of the barrel. There was no trigger guard, and the trigger ran half the length of the pistol grip to accommodate operation with a pressure suit glove. Altogether, it was a very groovy-looking (to use a period term) weapon that would have appealed to a nineteen-year-old moon soldier fresh out of space patrol boot camp.

SAUSAGE GUN: The next entrant in the space-age hand-held weapons race was the "sausage gun," a unique-looking little tube with nineteen tiny holes in the end, each a barrel for holding a tiny dart-like projectile. It even had a pen clip to retain it in a handy breast pocket of your space combat pressure suit. Each projectile was enclosed in a traditional-looking bullet casing, but the front part—which would normally be a projectile—was actually a "sabot," or protective covering that would fly apart once the dart was expelled from the gun, releasing the dart to impale its target. Interestingly, the sharp-nosed dart had fins on the back, clearly not needed in a vacuum, but perhaps this was intended for use once enemy soldiers had penetrated the moon base. This round was also propelled by traditional explosive powder, though the charge would have been tiny, at a speed of 4,000–5,000 fps. There was no hammer to trigger the explosive; it was electrically ignited via a pressure switch. It was semiautomatic in operation, and the darts would be spin stabilized for accuracy in a vacuum and presumably fin stabilized in an atmosphere. The "sausage gun" was six to eight inches long and about one and a half inches in diameter. Truly a tiny package for lunar death.

UNCLASSIFIED CONFIDENTIAL

SPIN STABILIZED MICRO GUN

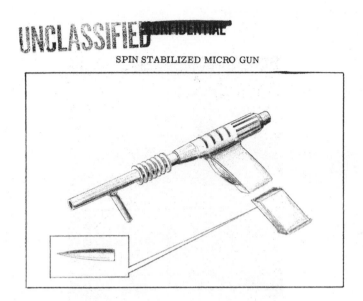

CHARACTERISTICS

```
Method of Propulsion . . . . . . . . . . .   Propellant
Projectile Weight. . . . . . . . . . . . .   .0027 lb.
Projectile Length. . . . . . . . . . . . .   .78 in.
Projectile Diameter. . . . . . . . . . . .   .14 in.
Muzzle Velocity. . . . . . . . . . . . . .   3000-4000 fps
Weapon Weight. . . . . . . . . . . . . . .   2-4 lbs.
Rate of fire . . . . . . . . . . . . . . .   Semiautomatic
Nr. of Rounds. . . . . . . . . . . . . . .   30-50
Weapon Length. . . . . . . . . . . . . . .   18-24 in.
Weapon Width . . . . . . . . . . . . . . .   1.5 in.
Weapon Height. . . . . . . . . . . . . . .   4-6 in.
```

UNCLASSIFIED CONFIDENTIAL

10

Fig. 20.1. This stylish pistol would fire a traditionally shaped bullet using a burning powder propellant like gunpowder (which can burn in a vacuum) to puncture an enemy combatant's space suit, wounding or possibly killing him. From DOD/US Army/NASA.

SAUSAGE GUN #2

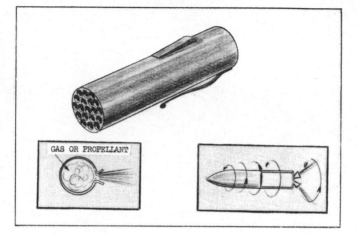

CHARACTERISTICS

```
Method of Propulsion. . . . . . . . Gas or Propellant
Projectile Weight . . . . . . . . . .005 lb.
Projectile Diameter . . . . . . . . 0.25 in.
Muzzle Velocity . . . . . . . . . . 3000 fps
Weapon Weight . . . . . . . . . . . 1 lb. or less
Nr. of Rounds . . . . . . . . . . . 19 to 37
Length. . . . . . . . . . . . . . . 6-8 in.
Diameter. . . . . . . . . . . . . . 1-1.5 in.
Method of Firing. . . . . . . . . . (Puncturing of Seal or
                                    (Ignition of Propellant
```

Fig. 20.2. This pistol would fire rocket-propelled pellets or darts that would puncture an enemy's space suit, incapacitating him instantly. From DOD/US Army/NASA.

SAUSAGE GUN #2: Similar in design to the above, this version fired pellets that had small nozzles on the back and contained a tiny amount of solid rocket propellant. When fired the propellant would ignite, giving the projectile added velocity and brief acceleration. Yet another variant had cylindrical darts, larger in diameter than those in the Spin Stabilized Micro Gun, with solid fuel charges and two angled nozzles on the back end to spin stabilize the projectile as it rocketed into a target.

DIRECTED GAS WEAPON FOR CLOSE-IN FIGHTING: This looked like the sausage gun but had a pistol grip attached. It had six barrels, and the "ammo" was a gas burst from an undefined explosive agent. The range of lethality was only three to six feet, so clearly this was intended as a last-ditch alternative to space suit–to–space suit close combat. It was five inches long and one and a half inches in diameter, and was in essence a handheld flamethrower.

GAS CARTRIDGE GUN: This was a single-barreled device that looked like a long-stem barbecue lighter. Along the top and bottom of the barrel were magazines that held small pellets .33 inches in diameter, about the same as a .32-caliber bullet. These pellets were lower velocity, about 1,000–1,500 fps, about the same as a powerful sporting air rifle. They were propelled by pressurized gas. The weapon was eight inches long by a half inch in diameter, perfect for hiding in your flight suit while making insincere peaceful overtures over at the enemy's lunar base just prior to assassinating the Soviet commander.

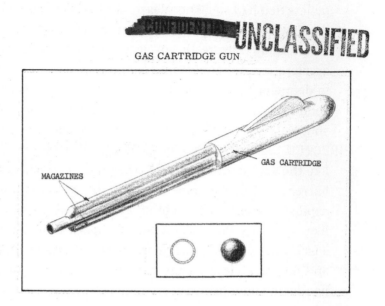

GAS CARTRIDGE GUN

CHARACTERISTICS

```
Method of Propulsion . . . . . . . . . . . . . .   Gas
Projectile Weight. . . . . . . . . . . . . . .   0.0012 lb.
Projectile Diameter. . . . . . . . . . . . . .   0.33 in.
Muzzle Velocity. . . . . . . . . . . . . . . .   1000-1500 fps
Weapon Weight. . . . . . . . . . . . . . . . .   2 lbs.
Nr. of Rounds. . . . . . . . . . . . . . . . .   25
Length . . . . . . . . . . . . . . . . . . . .   8 in.
Width. . . . . . . . . . . . . . . . . . . . .   0.5 in.
Height . . . . . . . . . . . . . . . . . . . .   3.5 in.
Pressure . . . . . . . . . . . . . . . . . . .   2000 psi
```

15

Fig. 20.3. This gas-powered weapon fired rocket-propelled pellets or darts that would puncture an enemy's space suit, which would be potentially fatal. From DOD/US Army/NASA.

SPRING PROPELLED SPHERICAL PROJECTILE: This pistol was another spacey-looking streamlined device, very handsome though less lethal than traditional handguns. It had a two-foot barrel about an inch and a half in diameter, tapering rapidly to the front. A curved pistol grip would cradle nicely in a space-suited hand. The ammunition would be another ball round, .20 inches in diameter or about the same size as a .22 bullet. This would be propelled by a spring much like your first Daisy BB gun. It held up to fifty rounds and projected the pellets at 1,000–1,500 fps. It would have been another short-range weapon.

GAS OPERATED NEEDLE GUN: Finally, we have the Gas Operated Needle Gun. Another Flash Gordon special, this weapon used pressurized gas to expend a tiny dart. A small magazine on the top held twenty-five darts. This was also a short-range weapon, with a dart velocity of 1,000–1,500 fps, and was sixteen inches long by two inches at the barrel.

While it is easy to mock these space-age Saturday night specials, it should be remembered that the object was not necessarily to kill your opponent, but merely to puncture his suit (with the possible exception of the Directed Gas Weapon for Close-In Fighting). This would not only cause immediate panic on the part of the soldier whose suit was breached by the weapon and now rapidly losing pressure, but would probably require the immediate assistance of one or more comrades to help him to safety back at the Soviet moon base, or at least to temporarily patch the leaking suit. This is a similar concept to many traditional combat weapons—their intention is to not merely kill enemy combatants, but to injure as many as possible to tie up other enemy soldiers who feel compelled to help them. Such is the cynicism of war.

Five pages of calculations follow the weapon descriptions, and indicate the expected diameter of suit penetration, the energy received by the victim when hit by a round, and the expected range of the weapons in the lighter gravity of the moon.

Recoil was minimized by most of these weapons by either the use of nontraditional propellants—gas and springs—or via smaller-than-traditional powder charges. Of course, even traditional firearms can be used in a vacuum to full effect, since the gunpowder contains its own oxidizer, so no atmospheric oxygen is needed for the ammunition to perform. One downside of lunar combat would be the smoke from the propellant—the exhaust would expand as a sphere at the end of the weapon, gradually reducing the visibility for the shooter. There is no wind to carry the smoky cloud away from the gun barrel.

There was also some concern about using high-powered projectiles. If not aimed properly, some specialists were concerned that the bullets could become tiny satellites, literally circling the moon and hitting you in the back—with no atmosphere and very light gravity, there would be little to alter its trajectory once fired at the proper angle. Of course, your aim would have to be perfect to accomplish this, and the charge would have to be a powerful one, about 5,500 fps—the fastest commercially available round is about 4,000 fps. But nobody wanted stray bullets circling the moon, being forgotten, and then hitting your own forces from the rear. It would have been the very definition of counterproductive.[4]

There are other concerns about using traditional firearms on the moon. The powder used to propel the bullets is somewhat temperature sensitive and, if too cold, might not burn well, resulting in a vastly lower bullet velocity. On the other end of the spectrum, if the gun got too hot a round might go off, but the generally agreed upon temperature for this to occur is over 350°F, so this is unlikely. One other concern, obliquely addressed in the paper, was that of like metals fusing. At the speed a projectile would be moving, it's possible, though unlikely, that this could occur inside a gun barrel. But coating the bullets with a substance dissimilar from the gun barrel would likely address this.

Of course, we've already looked at the US Army's Project Horizon and its proposed moon base, which included weapons from the army's inventory modified for lunar use (see chapter 2). These included Claymore mines, which would shoot hundreds of suit-puncturing pellets at enemy combatants, and the Davy Crockett M-29 nuclear howitzer, a bazooka-sized tripod-mounted defensive weapon that shot rocket-propelled, small-yield nuclear bombs. These had already both been tested (or, in the case of the Claymore, extensively fielded) on Earth and would likely have been effective on the moon if deployed there.

Our last entrant in the hand-held weapons race is called the Gyrojet, a rocket gun that was developed for use in Vietnam, not the moon. But it could have been used in an extraterrestrial setting. The Gyrojet was actually made and tested, and even sold commercially, though it was not considered a success. There was a pistol version and a carbine (rifle) version of the gun. It did not use a traditional explosive charge to propel the projectiles out of the gun, but used rocket power to fire the rounds. The ammunition was a .50-caliber rocket-propelled projectile, the same diameter as a modern heavy machine gun round.[5] When a cartridge was chambered out of the six-round-capacity magazine, and the trigger pulled, a plug of solid rocket fuel inside the cartridge was ignited. Four tiny nozzles—just small holes in the rear of the bullet—channeled the propellant in a slightly angled exhaust. The bullet left the gun under rocket power, spinning due to the off-axis exit ports, accelerating toward its target. It could reach a speed of about 1,250 fps. They were expensive and not very effective at short range, but at longer range the bullets built up enough velocity to be moderately destructive—in one demonstration, a Gyrojet round was reported to have blown a good-sized branch off a tree.

What about larger weapons for use closer to Earth? During the Ronald Reagan years, when the Strategic Defense Initiative (more commonly known as "Star Wars") was active, a lot of money was spent researching various space-based laser systems, to be powered via large volumes of highly reactive chemicals. The lasers were known as COIL weapons (Chemical Oxygen Iodine Lasers), and when the caustic chemicals were rapidly mixed, they could pump out a brief high-powered laser blast. The idea was to track and disable or destroy incoming enemy ICBMs before they could reenter the atmosphere and destroy American targets. Lots of research and testing was done, but none were used in space, due to various treaties and the difficulty of actually striking an enemy warhead within the complex physics of orbital trajectories.

One heavy-caliber space-based weapons system was actually tested, however. When the US was developing the Manned Orbiting Laboratory in the 1960s, the Soviet Union responded by designing its own military space station. When MOL was canceled in 1969, the Soviets had sufficient momentum to continue the development of their own space outpost. Three of these space stations were flown between 1973 and 1976, and were generally similar to, though smaller than, Skylab. These were the Almaz/Salyut space stations (see chapter 18). While arguably at least in part legitimate research stations, they were designed primarily for reconnaissance and other military purposes. And one of the Salyuts was armed. Salyut 3 carried a rapid-fire cannon called Rikhter for testing,[6] which fired up to 2,600 rounds per minute of 30mm (or possibly 23mm) ammunition.[7] Whatever the caliber, it reportedly carried only thirty-two rounds, so any test would be brief. It would be aimed by maneuvering the entire space station. Depending on which version of the murky post–Cold War USSR reporting one reads, the weapon was probably a modified tailgun from a Soviet Tu-22 bomber and was tested at least once. After the crew had departed the Salyut 3 station in mid-1974, the gun was fired via remote command—there were apparently concerns about the effects of noise and vibration of the station on the cosmonauts—and seemed to function properly. The bullets flew into space, joining the millions of other bits of space debris that circle the globe, but fortunately at an altitude too low to puncture the International Space Station or transiting Soyuz capsules.

The Almaz/Salyut stations were reportedly also designed to carry dual missile launchers, but these were apparently never tested in space.

When we think of space weapons, the mind naturally meanders to the lasers, phasers, blasters, and other directed-energy weapons of science fiction. But such systems have been difficult to scale down to hand-carried weapons of any power. The Rock Island Arsenal study summarized this in a brief comment that is as prescient today as it was in the 1960s: "The laser, for practical application as a weapon, is 20 years away."[8] They had the right idea but were off in terms of timing; such a weapon is still off in the future. But they will come in time; let's just hope that they are not carried into space.

CHAPTER 21

BURAN:
THE SOVIET UNION'S
ONE-FLIGHT WONDER

CLASSIFIED: *DECLASSIFIED AFTER THE FALL*
OF THE SOVIET UNION IN 1991

The shuttle sat on the launchpad, a white plume trailing from the huge fuel tank. It was the perfect night for a launch. The winged spaceplane hung from the side of the massive white fuel tank, poised for its first foray into space. A few miles away, controllers sat hunched over consoles, scanning numeric readouts and confirming, then reconfirming, values and cross-checks to make sure that this, the two-orbit maiden flight, would go off without a hitch. Then a plume of flame, and the beautiful white bird took flight, clearing the launchpad in seconds and sailing off into the night.

But something was not quite right. There was no Atlantic Ocean abutting the launch complex, just endless dark, dry fields. The gantry was oddly shaped, not quite what we were accustomed to seeing. Flight controllers spoke not in English, but in the crisp and abrupt tones of Russian. And the shuttle? With four side-mounted boosters and four more engines mounted on the bottom of where the external tank should be, not the three on the rear of the orbiter that we had seen before, it looked familiar, yet . . . wrong. The oddest part of all? There was not a single crew member aboard the shuttle, now soaring toward orbit. It was being flown robotically.

The date was November 15, 1988, the place Baikonur Cosmodrome in Kazakhstan. The spacecraft was the Soviet Union's answer to the US space shuttle, called *Buran* ("Snowstorm" or "Blizzard"). After a years-long crash program, the USSR had its own space shuttle in orbit, and it was performing well. Unfortunately, it would only fly once.

When plans for the US space shuttle began to circulate outside of NASA in the early 1970s (NASA, as a civilian agency, did not classify its programs), the Soviet military took notice. Some saw the shuttle for what it was—an expensive replacement for the Apollo moon rocket and space-craft. Others saw another purpose, which was also true—an attempt to create a militarily viable, quick-response spaceplane that could fly a wide variety of missions potentially detrimental to Soviet interests. While this second category of shuttle missions entered planning in the later stages of development (the US Air Force was brought in to the process to provide much-needed money just a few years before), it was by this time a parallel track to more prosaic civilian missions. And it is unlikely that the shuttle would never have been effective in a fast-response role.

Within a few years of the beginnings of full-scale shuttle devel-opment, the Soviets were working on designs for their own space-plane, headed up by aerospace designer Valentin Glushko, another member of the Soviet aerospace elite who had been responsible for designing their largest rocket engines. Like Eugen Sanger, he too had been designing rocket-powered aircraft since before WWII, but this program ended when Russia was invaded by Germany.

Despite his early experience in rocket planes, after considering many iterations for a modern shuttle in the 1970s, Glushko felt that NASA's design was well-founded, and would become the template for *Buran*. At the time the Soviets may have been unaware that the earlier designs were for smaller craft, with less wing and vertical stabilizer area, and that these had not been altered for flight considerations, but to accommodate air force needs. The USSR did, in any event, see the military potential in NASA's design, as well as military potential that was *not* there—some thought that the shuttle was intended to dive-bomb Russia with nuclear weapons. This was not a role to which it was suited, but nonetheless caused some Soviet generals to worry. As one Soviet space program expert wrote, "In 1976, despite apparent skepticism in the space industry, the Soviet government decided to respond to the 'Shuttle threat' with a similar spacecraft."[1]

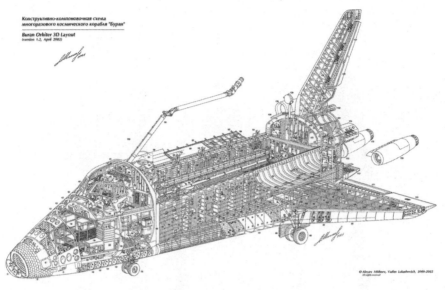

Fig. 21.1. A schematic view of the *Buran* orbiter. Image from www.buran.ru.

Specific concerns also focused on the US shuttle's large payload bay, sixty by fifteen feet, perfect for large military reconnaissance satellites (which was, in fact, what had driven the specification). The 50,000–60,000-pound payload capacity, and ability to also return cargo to Earth from orbit (a property unique, at the time, to the shuttle), was worrisome.[2] Returned cargoes could include captured Soviet satellites. And the large delta wings and tall rudder indicated a broad cross-range capability, another military specification. So they were understandably concerned. The Soviet analysts also later claimed (in what could have been some Monday-morning quarterbacking) that they did not see how the shuttle could be economically viable—there must have been something more to the program than was being advertised.

For these reasons and others, which may well have also included national pride, the Buran program was begun. It would be the most expensive space effort ever undertaken by the Soviet space program, with the lowest flight volume ever (just one test flight), as well as its last—the fall of the Soviet Union in 1991 signaled the end of its shuttle program as well.

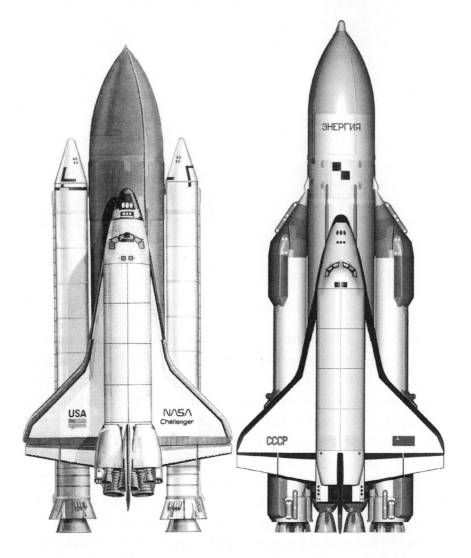

Fig. 21.2. When looking at the US space shuttle, to left, and the Soviet Union's *Buran*, to right, one might be excused for suspecting that some "involuntary information sharing" may have occurred. In fact, there were notable differences between the systems, but the basic engineering of *Buran*, which flew in 1988, appears to have had its origins in that of the shuttle, which first flew in 1981. Image from Roscosmos.

Externally, the *Buran* orbiter looks to be a dead ringer for the US space shuttle. But overall, *Buran*'s launch vehicle is completely different. It did not carry its own rocket engines for launch—those are thrown away along with the booster core. There were apparently plans to reuse the strap-on boosters, but since there was only one flight in the *Buran* configuration, this appears to not have been fully developed.

The booster core, what would be the external tank on the shuttle, is actually the Energia rocket. Designed as a multipurpose heavy booster, the Energia was a stand-alone system that was also capable of lifting *Buran*. The core rocket used four engines burning liquid hydrogen and liquid oxygen, just as the shuttle's main engines did. The four strap-on boosters had four engines each, using kerosene and liquid oxygen. Energia was an effective and powerful blend of the old and new. It was capable of lifting about two-thirds the payload of the Saturn V, or a bit more than 100 tons.

The orbiters were of almost identical gross measurements, but the similarity diverges after that. As mentioned, the orbiter did not carry any rocket engines for the boost phase onboard, so where the shuttle had three engines in back, *Buran* had only orbital maneuvering rockets. *Buran*'s payload bay was slightly larger than the shuttle's, and the wings, vertical stabilizer, and other primary structures were nearly identical.

The heat-shielding schemes were similar, with both orbiters covered in heat-resistant ceramic tiles (bottom, wings, and nose), reinforced carbon composite shapes (for areas facing extreme heat), and, on the later versions of the shuttle, heat-resistant fabric sheets. In both cases the tiles were a critical item, being fragile and prone to damage. But the glue used to affix the Soviet versions seemed to have worked better than the US mixture, without the same rate of loss during tests. Of course, much had been learned since the American program began with drop tests of the experimental shuttle glider *Enterprise* occurring in 1977, and the Soviets had a decade to benefit from their own research and testing. There have been reports of some damage to *Buran* after its only flight, but the true extent of this is unclear.

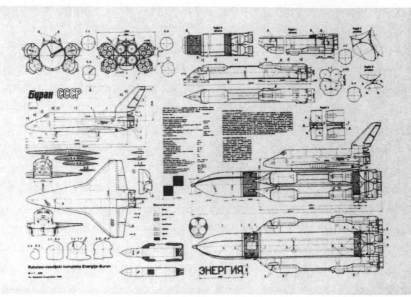

Fig. 21.3. A low-resolution copy of a general blueprint of the *Buran* launch system. Note the four liquid fueled boosters to bottom right, where the US shuttle had two solid rocket boosters. Image from Energia.

Much was unique in *Buran*, and it may in some ways have been an arguably superior design. It had a more powerful and more flexible launch system (the modular Energia booster), and a completely autonomous flight and guidance system. While shuttles had sophisticated guidance and avionics computers and software, they were not intended for unmanned flight. While many procedures and maneuvers could be accomplished autonomously, their software could not fly a complete mission. In fact, there were a number of studies written to explore the creation of a new software package to accomplish true unmanned shuttle flights for supply of the space station, but these were never fully implemented.

Buran on the other hand was intended from the beginning to be fully robotic if needed. This fit with a long history of automatic systems in Soviet spacecraft. While not always sophisticated, and wracked with trouble in the early years, by the 1970s the Soyuz spacecraft could reliably dock robotically, and in the late 1980s *Buran* would take this a step further with a completely automated flight profile, including the tricky landing.

Buran was the result of a long program of Soviet research toward the building of a spaceplane. Like the US, the Soviets had access to Sanger's *Silverbird* plans after the war, and these had spawned derivative designs in the USSR just as they had in the US. Vladimir Chelomei, the rocket designer who would struggle with Sergei Korolev for dominance of the Soviet space program, was a strong and early proponent of a rocket plane.

About the same time that plans for the US Dyna-Soar were getting serious, Chelomei was holding meetings with Soviet leadership about spaceplane programs. He had an especially sympathetic ear in Premier Khrushchev, though less so when Khrushchev was replaced by Brezhnev.

Chelomei's designs included a small test rocket plane to be ready for use in 1961, an unmanned satellite destroyer in 1962, and manned missions in 1963–1965. He also penned a draft of plans for nuclear spaceplanes that would fly well beyond Earth orbit, but these were apparently not seriously considered. One interesting difference between some of Chelomei's plans and those of others was the use of an ablative heatshield on his spaceplanes that would be jettisoned after its role in protecting the spacecraft had finished. Then the wings would fully extend, allowing the vehicle to land traditionally. It was an idea that would not require the extensive use of exotic alloys (such as the proposed X-15B and Dyna-Soar) or ceramic tiles (as on the space shuttle and *Buran*), but it was not fully developed. Two test flights of subscale rocket planes following Chelomei's designs were flown in 1961 and 1964, but after Khrushchev's removal from power, the program was terminated and Korolev was allowed to move ahead with his designs for Soyuz and the N-1 moon rocket (see chapter 14).

There was one smaller program of winged spaceflight for the USSR that preceded the Buran project, called *Spiral*. It was a tiny craft, only about one-twentieth the mass of *Buran*, started in 1965 as a response to both the Dyna-Soar project and various lifting body studies that

NASA and the air force were conducting. Unlike the unflown Dyna-Soar, *Spiral* flew a total of eight piloted tests, all subsonic, between 1967 and 1978. Unmanned subscale test versions were flown a number of times, perhaps as many as sixteen flights. The decision to proceed to the far more ambitious *Buran* resulted in the cancellation of *Spiral*.

Spiral, which was not much bigger than a fighter jet (but was bulkier), was intended for launch from a large air-breathing mothership. It had a single jet engine to give it once-around capabilities for landing, and would eventually have had a small payload bay similar to Dyna-Soar's. The wings were sharply swept, but had a variable angle ability and swung out to a wider configuration for lower-level flight and landing. It was a single-pilot spaceplane.

It is worth mentioning that Korolev had also worked on a rocket-powered glider, culminating with test flights in 1936; the advent of WWII ended this program. The Germans famously built and flew their own rocket-powered fighter planes in WWII that pressed that technology further, along with plans for manned V-2 variants.

Buran took its time going through the development and test processes, with over twelve years passing between its inception and its solo flight in 1988.

A problem for the Buran program, one that it shared at least in part with the shuttle, was finding appropriate missions to fly. While the shuttle was intended as a one-size-fits-all solution to America's space needs (a role that was greatly reduced after the loss of *Challenger*), *Buran* was a military program from the onset, designed to counter the "threat" of the shuttle. Various defensive and offensive schemes were developed for the spaceplane. An innovative rotary rack to hold nuclear missiles was designed for the payload bay. Laser systems and conventional missile launchers were considered. Orbiting space mine swarms were reportedly planned. Orbital reconnaissance packages were designed.[3] Other, less military tasks were also envisioned, such as carrying modules for the planned assembly of the Mir-2 space station (which was eventually incorporated into the Russian component of the International Space Station).

The November 15, 1988, mission was the only flight of the entire program. *Buran* boosted into orbit, shedding first the strap-on boosters, and eventually the main Energia booster stage, as it flew. After two orbits and 53,000 miles, the spaceplane reentered and landed back in Kazakhstan under autonomous control. It was the shortest successful manned Soviet flight since Yuri Gagarin's single orbit in 1961.

The orbiter was eventually moved, along with another airframe used for testing and research, into a hangar not far from the launch facility. And there it sat until May 2002, when the poorly maintained hangar's roof collapsed, killing a number of workers.[4] *Buran* was crushed.

The Buran program was initiated by a variety of factors, including poor military intelligence, competitiveness, and nationalism. On the other hand, the program planned and built a well-designed rocket booster and spaceplane that could have been the backbone of a successful program that would have allowed the Energia rocket—along with the Soyuz and Proton rockets—to perform missions appropriate to their capabilities, and *Buran* could have flown missions best suited to it.

The end came almost as swiftly as the demise of the Soviet Union, though the termination of *Buran* was not officially announced until 1993. What could have been an enabler of sensational space stations and an increasingly vibrant manned orbital program became just one more episode in the long string of canceled and underdeveloped spaceflight initiatives born of the space age. The Soyuz system carried on in its stead.

As a historical footnote, documents obtained by the Freedom of Information Act have generated claims that the resemblance between *Buran* and the US shuttle was based on more than converging lines of research and "form following function." A number of journalists have written on extensive and costly efforts by the Soviet Union to gain information on the US shuttle beyond the already massive amounts available to the general public. While there was an enormous amount of technical material about the shuttle published by NASA and available from the contractors—for this, one would not have needed to engage in espionage or skullduggery—areas specific to US Air Force functions were classified and some other elements restricted, so there was more information to be mined from less legitimate sources.

A report drafted in 1985 but only declassified within the last six years puts it plainly in the preface: "In recent years, the United States Government has learned of a massive, well-organized campaign by the Soviet Union to acquire Western technology illegally and legally for its weapons and military equipment projects. Each year Moscow receives thousands of pieces of Western equipment and many tens of thousands of unclassified, classified, and proprietary documents as part of this campaign. Virtually every Soviet military research project—well over 4,000 each year in the late 1970s and over 5,000 in the early 1980s— benefits from these technical documents and hardware. The assimilation of Western technology is so broad that the United States and other Western nations are thus subsidizing the Soviet military buildup."[5]

There were clear indications, the intelligence agencies felt, that the Soviets were using more than the publicly available information on the shuttle. Inside the report, it was claimed,

> [F]rom the mid-1970s to the early 1980s, NASA documents and NASA-funded contractor studies provided the Soviets with their most important source of unclassified material in the aerospace area. Soviet interests in NASA activities focused on virtually all aspects of the space shuttle. Documents acquired dealt with airframe designs (including computer programs on design analysis), materials, flight computer systems, and propulsion systems. This information allowed Soviet military industries to save years of scientific research and testing time as well as millions of rubles as they developed their own very similar space shuttle vehicle.[6]

Regarding specific bits of black-market acquisition: "Requirements for documents alone can command amounts as considerable as hardware; examples include over 50,000 rubles (roughly $140,000 in 1980 purchase power equivalents) for documents on the US shuttle orbiter control system . . ."[7] The report goes on to cite examples of public database plundering, and efforts to gather additional information from major research universities in the US, including Stanford, UC Berkeley, MIT, Caltech, and many more.

In response, President Reagan allegedly approved a CIA plan to deal with the problem in a particularly creative way. The spy agency concocted believable but deliberately flawed data to feed Soviet agents and information collection efforts, as well as misleading descriptions of legitimate shuttle systems, such as placement of critical heat-shielding tiles. Sure enough, according to another journalist, the dangerously erroneous tile placement showed up on *Buran*.[8] There are even reports of "modified" components being made available to collectors that the CIA was fairly certain would make their way back into the hands of Soviet engineers for evaluation and reverse-engineering—note this excerpt from Gus Weiss of the National Security Council: "I met with William Casey, Reagan's Director of Central Intelligence, on a frosty afternoon in January 1982. . . . I proposed using the Farewell [the code name of the CIA operation] material to feed or play back the products sought by [the Soviets], only these would come from our own sources and would have been 'improved,' that is designed so that on arrival in the Soviet Union they would appear genuine but would later fail."[9] These apparently included modified or degraded heat-shielding material samples.

We may never know how much of the *Buran* effort was original Soviet engineering and how much was "borrowed" from NASA's shuttle designs and engineering. The similarities are striking but imperfect. And the Buran project, while it ran for well over a decade, was quite brief operationally. Both spacecraft had their strong points and weak points, but the US shuttle, for its faults, was a workhorse well into the twenty-first century. *Buran* might have done the same, but many had their doubts. One such view on the Buran program came from Gherman Titov, the second cosmonaut to fly into space, and may in the end best summarize the Soviet spaceplane: "It would be better if we just cut it up for scrap . . . for that is all that it is worth."[10]

CHAPTER 22

MAJOR MATT MASON: A MAN FOR THE NEW SPACE AGE

UNCLASSIFIED

If there was a toy that was iconic of the space age, it was the Major Matt Mason line, which the Mattel toy company debuted in 1966. There were others. . . . Hasbro's GI Joe featured a Mercury capsule and a silver space suit for the scar-faced soldier, many companies made tiny plastic figures of "moon soldiers," and much later the retro Buzz Lightyear of *Toy Story* fame hit the silver screen and retail shelves in a big way. But it was the range of Matt Mason toys that marched to the drumbeat of the American space program in the mid-1960s.

What place does this have in a book about unusual programs of the space age? Simply this: the toy line took the basic premise of the lunar landing program as it was in the early 1960s and created a line of products based loosely upon it, expanding into unusual and, later, bizarre designs that echoed the equally unusual projects littering the drawing boards of aerospace firms and NASA itself. To a generation of young boys (and fortunate girls), Matt Mason became emblematic of NASA's great voyages of lunar exploration.

The toy line was initiated with a six-inch-high astronaut named Matt Mason, a major in an undefined space service, who was billed as "Mattel's Man in Space." The space-suited figure was made of a rubberlike plastic formed over wires, which allowed it to be posed in a broad range of positions. A hard plastic helmet with a flip-up visor completed the early action figure.

Accessories were loosely patterned after NASA project designs. There was a jet pack (somewhat akin to the US Air Force's AMU), a "Moon Sled" that the astronaut rode standing up, and a "Space Crawler," which was an ungainly lunar transport device—battery powered—that had spidery, rotary legs instead of wheels.

As the toy line diversified, more astronauts were added with color-coded space suits, including an African American astronaut, at a time when NASA's flying corps was far from integrated,[1] either racially or in terms of gender (Mattel's Matt Mason toy line remained "men in space" only; however, Mattel's Barbie had her own "Miss Astronaut" outfit in 1965, beating the major by a year). There was also a Matt Mason lunar hard suit that the figures could be put into; it was an almost comical-looking arrangement that looked like an enormous trash can with arms, legs, and windows. But it was based on a proto-type lunar hard suit designed by a JPL engineer, and therefore had a vaguely legitimate place in the lineup.

The pinnacle product was the two-foot-high space station, which was actually a lunar outpost with three stories, two of which were "pres-surized" (i.e., they had transparent flip-down plastic panels that allowed users to reach inside and manipulate the toy figures). The floors were metal gratings, very similar to what would later appear in Skylab.

As the product line's short life progressed, the direction moved quickly from legitimate space age themes to science fiction ones. There had always been a laser rifle (suspiciously close to a design from the television series *Lost in Space*), but increasingly the products became more weaponized and fantasy-oriented. Foam-wheeled lunar hot rods, a single-tracked tank-like rover, and a giant laser cannon were included in the lineup. The toys were imaginative, but had left the realm of plausibility.

The toy line faded after 1970. In a further parallel to the space race, Mattel saw the same writing on the wall that NASA did; public support of space exploration was on the wane. Mattel shifted to more prosaic toys, and the Matt Mason line was quietly dropped to make way for higher-volume products.

At the same time, NASA moved into the space shuttle age, abandoning its hard-won Apollo hardware, and the air force accepted the inevitable superiority of robots for orbital espionage and dropped its manned spaceflight ambitions for the time being.

In a perhaps imperfect metaphor, the interest of the average American adult in space exploration waned at about the same time that American children turned away from realistic space toys—NASA had made space exploration look too easy. Astronauts had been to the moon, and the shuttle was perceived as a boring space truck. Children likewise no longer wanted to play with astronaut dolls that mimicked what NASA was developing for the exploration of the moon.

The almost heroic competition and excitement of the space race had ebbed, replaced by routine orbital ventures, a number of Earth-orbiting space stations, and the robotic exploration of the planets. All this had its appeal, but was by no measure as popular as the early lunar expeditions had been with kids or grown-ups.

And what of the future? Will public fascination with spaceflight be rekindled as NASA and SpaceX reach for Mars, as Virgin Galactic sends tourists into space, and orbital hotels are regularly visited by Jeff Bezos's Blue Origin rockets? Will a new line of space toys and electronic games focus on Martian expeditions and space colonies? The private ventures mentioned are moving ahead as quickly as they can, and NASA continues to refine its plan to get astronauts to Mars by the mid-2030s. Roscosmos, the Russian space agency, is planning similar ventures closer to Earth. The European Space Agency, which has flown its own astronauts on both US and Russian spacecraft, is discussing an ambitious international "lunar village."[2] China plans to send humans to the moon within a decade and is aggressively pursuing its own orbital space program, including space stations, the second of which was launched in 2016. India has orbited an unpiloted spacecraft around Mars and has human spaceflight ambitions. And as mentioned, private enterprise has entered the new space race in a big way. While aerospace companies have always built NASA's hardware, these were created via paid contracts from the US government. The

new landscape is somewhat different and involves massive invest-
ments of private capital.

Preeminent among the new players is SpaceX, Elon Musk's rocket
company. Started just over a decade ago with his portion of the proceeds from the sale of PayPal, Musk has created a company that now
launches NASA, air force, and commercial payloads with stunning regularity. Despite the inevitable setbacks, his success rate is admirable,
and his plans audacious. For starters, SpaceX has been successful in
launching rockets with large numbers of clustered engines—something the Russians had worked with for decades, but which was not
NASA's preferred method. The upper limit for liquid engines on NASA
rockets has to date been the Saturn V with five engines on the first stage;
SpaceX's Falcon 9 uses nine engines, and the Falcon Heavy rocket will
use twenty-seven on the first stage alone.[3] The Falcon Heavy is poised
to become a new leader in heavy-lift capability for the US when it flies
in 2017 or 2018 (followed by NASA's SLS heavy booster soon thereafter). And SpaceX's rockets are designed to fly back to their launch
points (or nearby sea-going barges) for refurbishment and reuse,
which should finally fulfill the vision of the early space shuttle designs
for economical and frequent spaceflight on a budget.

Jeff Bezos of Amazon fame has followed a somewhat similar path,
but is serving both a higher and lower (no pun intended) segment of
the marketplace. His Blue Origin rockets have flown numerous times
to suborbital altitudes, testing both reusable boosters and capsules
similar to SpaceX's. These are the core of his tourist-driven suborbital
rocket business. On the other end of the spectrum, he has a contract
in hand to design the new high-power engines for United Launch Alliance's (ULA) replacement for the Atlas rocket, which, until SpaceX
came into the market, dominated (along with the Delta II) US launch
services. The Atlas dates back to the 1950s, and was extensively redesigned to use rocket engines sourced from Russia, reentering the
market in 2002 after an operational hiatus. But with increasingly tense
relations between Russia and the US, a new first-stage rocket engine
design was solicited, and Bezos's company seems to have the job. The

new engine, the BE-4, will be used to power ULA's replacement for the Atlas, called the Vulcan.

Richard Branson's Virgin Galactic continues on its quest to build the first tourist spaceflight vehicle. Carried aloft for launch by a larger plane in a fashion similar to the X-15, SpaceShipOne, and its updated version, SpaceShipTwo, has been extensively tested. Despite a catastrophic crash in 2014,[4] the company continues to develop the spaceplane and expects revenue flights to begin before 2020. At the current time, increased effort is being directed at a satellite delivery system based on similar technology—unpiloted spacecraft have much lower safety require-ments, and are viewed by Virgin as a good parallel business.

Robotic mining in space is another area generating increasing investment. A wide variety of plans and schemes have come to the fore in recent years, with companies such as Planetary Resources and Deep Space Industries actively designing robotic spacecraft to mine aster-oids with technology that would have been pure science fiction just a decade ago. Lunar mining is being pursued, with Moon Express at the forefront of these efforts. Hardware has been built and may fly within the next couple of years. The commercial potential of these efforts should be vast—it is expensive to launch water and metals into orbit, and the ability to extract these resources from asteroids and the moon, if achievable, could be a game changer.

And the satellite industry continues to grow at a rapid pace. Since the 1960s, the telecommunications industry has been the linchpin of Earth-orbital operations, and hundreds of companies are reaping bil-lions of dollars of revenues in this area. The advent of microsats and nanosats—small satellites utilizing modern technology derived, in many cases, from the guts of smartphones and their brethren—have reduced the cost of developing and launching these useful and profit-able devices. They are becoming a major market force.

We are on the cusp of a new space age, with a seemingly limitless opportunity for both robotic and human engagement in space. How this will progress is a combination of national will and market forces. All of these efforts build on the heavy lifting accomplished during the space race, with most of the basic problems already solved by government-employed engineers in both the US and Russia. The new programs discussed here have all drawn upon, and built upon, these accomplishments, which are available for anyone to capitalize on.

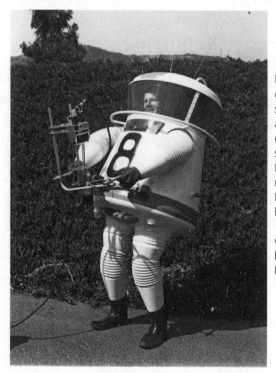

Fig. 22.1. An early design for a hard moon suit, which was also an inspiration for the earliest Matt Mason space action figures. Designed by a JPL engineer named Allyn "Hap" Hazard (which is an even better moniker than "Matt Mason"), the suit was never developed by NASA. Image from NASA/JPL–Caltech.

In this brave new world of space, many oddball designs and plans will come and go, just as they did in the 1950s and 1960s. There will be dead ends, bankruptcies, and restarts aplenty, but the forward momentum will continue what the national space agencies started.

And what of those space agencies? If the pundits have their way, NASA and perhaps Roscosmos, Russia's space agency, will increasingly provide seed funding and support for commercial space enterprises, as the agencies themselves refocus on pure exploration and ventures beyond low Earth orbit. The new space age appears to be a place of great promise.

Fig. 22.2. SpaceX's Falcon Heavy, a high-power booster that is set to revolutionize access to space in 2018. The most powerful American rocket since NASA's Saturn V of the 1960s, the Falcon Heavy will be capable of delivering about 120,000 pounds to low Earth orbit (roughly half the capability of the Saturn V, but double that of the space shuttle). Image from SpaceX.

In a reflection of this trend, even Major Matt Mason has a new lease on life. Companies have popped up to provide replacement parts and reproductions of the now-collectable toy figures, and a new release of the space-age toy has been discussed. There is even a movie in the works under the steady hand of Tom Hanks, who was guided to his role in *Apollo 13* and the production of *From the Earth to the Moon* by his own childhood time spent with Matt Mason toys.[5] And the irrepressible Buzz Lightyear lives on in toy stores and Disney theme parks worldwide.

After decades of slow growth and frustrated attempts to launch new space initiatives, it seems that the future for both humans in space and space-suited action figures is bright. We will soon have a rich new collection of amazing stories to tell as our children unwrap their toys, based on the new space age, on Christmas morning.

ACKNOWLEDGMENTS

Any book of this kind benefits greatly from the input of experts from the field, and this one is no exception. A number of kind individuals gave generously of their time to proofread, fact check, and offer advice to improve the content immeasurably. These include:

Michelle Evans, John B. Charles, David Clow, David Hitt, Jay Chladek, Rachel Tillman, Pascal Lee, and especially Susan Holden Martin and Francis French, who both surely became far more intimate with the book than they ever envisioned. My deepest thanks.

Leonard David, Andy Chaikin, Peter Orton, Rob Kirk, Steve Fentress, Pat Kilbane, Jeff Kanipe, and Melanie Melton Knocke all provided thoughtful supporting blurbs for the book.

The wonderful crew at Prometheus books went above and beyond, as always, to create a fine product. Steven L. Mitchell approved a book that was a passion project for me, and I am in his debt. There are many others with Prometheus who deserve praise, but top nods go to Jeffrey Curry, a well-informed and extremely patient editor, who left the book many times better than he found it. Hanna Etu walked me through the often tortured landscape of image and photo clearances with grace and good cheer. Further thanks are due to Jill Maxick, Catherine Roberts-Abel, Nicole Sommer-Lecht, and Bruce Carle.

John Willig of Literary Services Inc., agent and friend, was right where an author needs him to be at every turn. He is, quite simply, the steely-eyed missile man of nonfiction agents (in this genre, that's a huge compliment).

Many fine writers are due credit for the deep and laborious research they performed for their own works, from which I learned much. These include a number of names from above, and also: David Portree, Dwayne Day, Jeff Foust, Wernher von Braun, David Scott, Buzz

Aldrin, Boris Chertok, Steven Dick, Peter Merlin, Frederick Ordway III, Anatoly Zak, Edward Clinton Ezell, Linda Neuman Ezell, William Corliss, and the many other uncredited but talented and devoted individuals who have written material for NASA, the various US security agencies, and general journalism on space subjects for the past six decades. My thanks to you all.

And finally, a nod to Gloria Lum: please live long and prosper. The world needs your art, and I need your friendship.

Any remaining omissions, oversights, or misstatement of facts are my responsibility.

NOTES

CHAPTER 1: NAZIS IN SPACE: PROJECT SILVERBIRD

1. Chris Gainor, *To a Distant Day: The Rocket Pioneers* (Lincoln, NE: University of Nebraska Press, 2008).

2. Gregory P. Kennedy, *Vengeance Weapon 2: The V-2 Guided Missile* (Washington, DC: Smithsonian Institution, 1983).

3. James P. Duffy, *Target: America: Hitler's Plan to Attack the United States* (Westport, CT: Praeger, 2004).

4. Eugen Sanger and Irene Sanger-Bredt, *A Rocket Drive for Long Range Bombers*, trans. M. Hamermesh (Cambridge, MA: Radio Research Laboratory, 1952). Translation was done in 1952; when the German original was written in 1944 these records were still classified within the German military.

5. At one point Sanger suggests the advantages of using a rocket engine capable of *ten times* this thrust, or about twice the initial rating of early versions of the Saturn V's F-1 engine: "While the 100-ton rocket motor weighs 2500 kg, one must assume a weight five times as great (i.e., 12,500 kg) for a well-regulated motor with 10 times greater maximum thrust." Ibid.

6. Ibid.

7. This quote and those that follow from Sanger sourced from: Sanger and Sanger-Bredt, *Rocket Drive for Long Range Bombers*.

8. This figure is stated at differing values in various accounts. In the 1944 paper, Sanger suggests a potential payload mass of thirty *tons*.

9. In April 1942, sixteen American medium bombers took flight from an aircraft carrier operating near Japan. They headed for the mainland and made their bomb runs, scattering their 500-pound explosive payloads over military and industrial targets across a number of cities. Unable to land on the aircraft carrier, the planes attempted to reach China with varying degrees of success. The physical damage to Japan's war effort was negligible, but the psychological effect on the country was immense. Historians have speculated that the refocusing of Japan's military on defenses against enemy aerial sneak attacks was a factor in that country's loss of WWII.

10. Some American and Allied freighters were torpedoed within sight of US shores on the East Coast early in the war, but this was not generally perceived as a threat to home and hearth.

11. Claus Reuter, *The V2 and the German, Russian and American Rocket Program* (Canada: German-Canadian Museum, 2000), p. 87.

12. NASA, Columbia Accident Investigation Board, public hearing transcripts, 2003.

CHAPTER 2: RED MOON: COUNTERING THE COMMUNIST THREAT ON EARTH AND IN SPACE

1. Michael Neufeld, *Von Braun Dreamer of Space, Engineer of War* (New York: Alfred A. Knopf, 2007).

2. Ibid.

3. Ibid.

4. Ibid.

5. Roger Bilstein, *Stages to Saturn, a Technological History of the Apollo/ Saturn Launch Vehicles* (Washington, DC: NASA, 2013).

6. William J. Jorden, "Soviet Fires Earth Satellite into Space," *New York Times*, October 5, 1957, p. 1.

7. Constance McLauchlan Green and Milton Lomask, "Vanguard, a History," NASA, Washington, DC, 1971.

8. "What Flopnik! Britons Blare," *Chicago Tribune*, December 7, 1957, http://archives.chicagotribune.com/1957/12/07/page/1/article/what-flopnik-britons-blare; Suzanne Deffree, "1st US Satellite Attempt Fails, December 6, 1957," *EDN Network*, December 6, 2015, http://www.edn.com/electronics-blogs/edn-moments/4402889/1st-US-satellite-attempt-fails--December-6--1957 (accessed August 12, 2016).

9. Neufeld, *Von Braun: Dreamer of Space*.

10. *Project Horizon Report: Volume I*, June 9, 1959, United States Army.

11. This and further references to the Horizon study: *Project Horizon Report: Volume I, Summary and Supporting Considerations*, June 9, 1959, United States Army.

12. I've added the italics, because, while they could not have known

it at the time, much current interest in lunar exploration and development is derived from the realization there are resources such as water there, in the form of ice near the poles, from which rocket fuel and consumables for life support can be derived. The lunar soil also contains metals from which hardware may one day be manufactured.

13. It's worth noting that the 1.5-million-pound thrust engine called the F-1—five of which would power each of the Apollo moon voyages—had already been commissioned by the US Air Force, but would not be ready for six years.

14. Depending on how you count them, there may have been as many as twenty-two attempted launches of the smaller Saturn flown. This includes the Saturn 1 described here, as well as the later and improved Saturn IB, which had a more powerful upper stage. See "Saturn Illustrated Chronology," Appendix H, *Moonport*, Appendix A, *Apollo Program Summary Report*, Appendix A.

15. "Project Apollo: A Retrospective Analysis," http://history.nasa.gov/Apollomon/Apollo.html#note36 (accessed August 12, 2016).

16. Marsha Freeman, *How We Got to the Moon* (includes an interview with Harry Ruppe from December 1992 in Munich) (Washington, DC: 21st Century Associates, 1993).

17. Larry Grupp, *Claymore Mines: Their History and Development* (Boulder, CO: Paladin, 1993).

18. Phrases such as "manned spaceflight," "first man in space," and other gender-related terms were specific to the period under discussion, and are in no way intended to imply the views of NASA, the aerospace contractors, or myself with regard to equal opportunities in spaceflight, engineering and the sciences, or other pursuits. Such dated terminology remains in the text to relate the spirit of the times, and when referring to modern endeavors, every effort was made to update the language used accordingly. Many women and minorities have contributed immensely to spaceflight and the exploration of our cosmos, and their accomplishments have made our world a far better place.

19. Matthew Johnson and Nick Stevens, *N-1: For the Moon and Mars: A Reference Guide to the Soviet Superbooster* (Livermore, CA: ARA, 2014).

CHAPTER 3: *DAS MARSPROJEKT*: RED PLANET ARMADA

1. Wernher Von Braun, *Das Marsprojekt* [*The Mars Project*], trans. Henry J. White (University of Illinois Press, 1953).

2. The Arizona Territory was created in 1863, and did not become a state until 1912.

3. Refracting telescopes use lenses as the primary light-gathering optic, as opposed to the mirrors used in most modern telescopes. They can be fine instruments, but are generally smaller than a reflecting instrument of equivalent cost.

4. Burroughs wrote a series of fictional accounts about Mars between 1911 and about 1950, published both in pulp magazines and as books. In these tales, an American Civil War veteran named John Carter travels to the red planet via a sort of teleportation, and is quickly embroiled in wide-ranging struggles to protect the interests and honor of his beloved Dejah Thoris, the princess of Mars.

5. Percival Lowell, *Mars* (Boston: Houghton Mifflin, 1895).

6. Ibid.

7. Ibid.

8. It was not until the Soviet Union began selecting cosmonauts in the late 1950s that the idea of sending women into space, as well as men, became a practical consideration. Valentina Tereskova became the first woman sent into space in 1963 aboard Vostok 6.

9. The *El Paso Times* wrote up the incident on May 30, 1947: "El Paso and Juarez were rocked Thursday night when a runaway German V-2 rocket fired from the White Sands Proving Ground in New Mexico crashed and exploded on top of a rocky knoll three and a half miles south of the Juarez business district. The giant missile burst in a desolate area of jagged hill, gullies and boondock. No one was injured. Lt. Col. Harold R. Turner, White Sands commanding officer, said failure of the rocket's German-made gyroscope caused it to swerve from its set northerly course." Lady Luck was truly with the German rocketeer.

10. Von Braun, *Das Marsprojekt*, p. 1.

11. Ibid., p. 3.

12. Note the word "logistical." While launch facilities closer to the equator are more efficient locations from which to launch orbiting rockets, getting the

needed hardware and fuel supplies to Johnston Atoll would have been far more challenging than sending them to Florida. See paragraph details.

Notably, Jules Verne also chose this general area for the launch of his moonship, the *Columbiad*, in his landmark novel *From the Earth to the Moon*, published in 1867. "Launch" may be a misleading term, as the bullet-shaped craft was, in the book, fired from a giant cannon.

13. "Saturn Illustrated Chronology—Part 7," NASA, http://history.nasa.gov/MHR-5/part-7.htm (accessed September 11, 2016).

14. In a preface to a 1962 edition of the book, von Braun pointed out that with the newer fuels being used on his Saturn rockets—liquid hydrogen and liquid oxygen—the thrust generated was about half again as much as the hydrazine and nitric acid combination. So, at least for the Earth-to-orbit space ferries, the gains would have been substantial—a two-stage rocket of modern design would carry about five times the mass into orbit as his earlier design. This would have greatly reduced the size of the ferry, the amount of fuel to be handled, and the number of trips it would have to make. He further pointed out that the advances made in nuclear rocket engines could have doubled that efficiency, which would have been a boon to the interplanetary leg of the journey.

15. The cryogenic technology used over long durations in the Apollo program were for storing liquid oxygen for generating power and breathable oxygen during the trip to the moon and back. The propellants used for the trans-lunar journey were hypergolic, i.e. Aerozine 50 (a form of hydrazine) and nitrogen tetroxide.

16. As noted, the atmospheric pressure on Mars is well less than one one-hundredth that of Earth; von Braun had it pegged at about 50 percent.

17. Edgar Rice Burroughs, *Warlord of Mars* (Chicago: A. C. McClurg, 1919).

18. Pascal Lee, in a written statement to the author, 2016.

CHAPTER 4: PROJECT ORION: WE COME IN PEACE (WITH NUCLEAR BOMBS!)

1. Actually, that last part is not entirely accurate. There was often some Brooklyn-accented, lower-rank, wise-cracking private along for comic relief. He was usually a last-minute addition to the crew, perhaps a trusted mechanic

with a foot-long wrench in his hip pocket. In at least one film, *Robinson Crusoe on Mars*, the role was filled by a space-suited monkey.

2. Fortunately for some of us who are fans of this bygone genre, many of these films (the more prominent of them anyway; hundreds of B-grade spinoffs were made that were far worse and less ambitious) were based on some classic science fiction novels of the era, penned by greats like Robert Heinlein. The underlying stories, if not the final performances, were often compelling.

3. C. J. Everett and S. M. Ulam, "On a Method of Propulsion of Projectiles by Means of External Nuclear Explosions," *Los Alamos National Laboratory Report LAMS-1955* (August 1955).

4. Ibid.

5. George Dyson, *Project Orion: The True Story of the Atomic Spaceship* (New York: Henry Holt, 2002). The tonnage of the largest designs is somewhat unclear, ranging from a claimed 40 million tons to, in some accounts, 100 million. Suffice it to say it was really, really big. Most of the study focused on smaller variations.

6. Freeman Dyson, *Disturbing the Universe* (New York: Harper, 1979), pp. 109–10.

7. G. R. Schmidt, J. A. Bonometti, and P. J. Morton, "Nuclear Pulse Propulsion—Orion and Beyond," 2000, pp. 4–5, http://ntrs.nasa.gov/archive/nasa/casi.ntrs.nasa.gov/20000096503.pdf (accessed August 30, 2016).

8. "National Average Wage Index," Social Security, https://www.ssa.gov/OACT/COLA/AWI.html (accessed August 30, 2016).

9. John McPhee, *The Curve of Binding Energy* (New York: Farrar, Straus and Giroux, 1974), pp. 167–68. At the time, nothing was known for certain about interplanetary radiation, the bane of human Mars mission planners. However, given the incredible lifting power of Orion, the problem would probably have been solved with mass.

10. Dyson, *Disturbing the Universe*, p. 115.

11. The air force did, however, become the leader in unmanned systems such as satellites, ICBMs, and, now, long-duration robotic spaceplanes.

12. To be more specific, specific impulse equals the change in momentum of an object delivered by unit of propellant used. For a concise explanation by NASA, see "Specific Impulse," NASA, https://www.grc.nasa.gov/www/k-12/airplane/specimp.html (accessed September 4, 2016).

13. "Saturn Illustrated Chronology—Part 7," NASA, http://history.nasa.gov/MHR-5/part-7.htm (accessed September 11, 2016).

NOTES **307**

14. Eugene Mallove and Gregory Matloff, *The Starflight Handbook* (New York: John Wiley and Sons, 1989), p. 60.

15. By the advent of the Viking/Voyager missions, the press materials were a bit more open about the use of plutonium in spacecraft power supplies. But during the Apollo program, when plutonium was used to power experiments taken to the moon, there was far less fanfare.

16. Project Orion lives on in more recent studies. One example is a Sandia National Labs study by Johndale Solem, "Some New Ideas for Nuclear Explosive Spacecraft Propulsion," LA-12189-MS, October 1991.

CHAPTER 5: LUNEX: EARTH IN THE CROSSHAIRS

1. US Air Force, "Lunar Expedition Plan: LUNEX," USAF WDLAR-S-458, May 1961.

2. Emphasis by the author. It is oddly refreshing to see the study come right out and say it: we want to hurl nukes at the USSR from the moon.

3. As NASA learned with the space shuttle, reusable space vehicles are harder to engineer than assumed in studies like LUNEX and *The Mars Project*. Commercial companies like SpaceX and Jeff Bezos's Blue Origin continue to explore reusable spacecraft today, with vastly improved technologies. It is still hard to accomplish.

4. As an example, one 1965 report cites an estimated reliability of 96 percent. "Apollo Reliability and Quality Assurance Program Quarterly Status Report, Second Quarter, 1965," NASA Office of Manned Spaceflight, N79-76650. This was later estimated to be closer to "three nines," or 99.9 percent.

5. It can be argued that this capability would have been useful for the kind of inflight emergency that Apollo 13 suffered, when an oxygen tank on the service module, the back half of the command module, exploded on the outward leg of the journey to the moon. But rendezvous would have been impossible, and without the capability of the separate lunar module to house the crew during the loop around the moon and return to Earth, the crew would have been dead long before any hope of rescue. Later, in the Skylab program and the last half of the shuttle program, backup rockets and capsules were sometimes kept on ready standby for rescue missions when the planning indicated that a rescue could be feasible.

6. Linda Neuman Ezell, *NASA Historical Data Book, Volume II*, 1988, p. 121, http://history.nasa.gov/SP-4012v2.pdf (accessed August 31, 2016).

7. There would have likely been some cost savings using military personnel instead of higher-paid civilians to create the US Air Force lunar landing program. However, civilian contractors would still have been needed, likely moving the programs closer to parity. And the demands of testing a reentry-capable lifting body and all the other associated hardware and technical capabilities needed to accomplish the program would have mounted quickly—nobody knew at the time how difficult spaceflight, even just to Earth orbit, would be. In the end, LUNEX may have cost less and shaved a year or two off Apollo's schedule, but it is doubtful it would have amounted to a vast difference. It's also possible that it would have ended up being sidelined, or at least delayed and saddled with a ballooning budget, due to the large number of technical challenges involved.

8. One of Phillips's few (and involuntary) high-profile moments was when a report he had quietly compiled, critical of one of the primary Apollo contractors, came to light during congressional hearings convened after the Apollo 1 fire in 1967. Phillips was no fan of the contractor, North American Aviation (NAA), which was building the Apollo command and service module as well as the Saturn V's second stage. This unflattering report, authored in 1965, was sent to the president of NAA as well as to top NASA brass. He also gave his NASA colleagues a verbal briefing. The observations and suggestions within were not acted on in a responsible manner—some of which could have resulted in a safer spacecraft, possibly averting the Apollo 1 accident that killed three of NASA's moon-bound astronauts. In the end, that accident resulted in the implementation of many of his recommendations, and then some. The NAA command and service modules were among some of Apollo's most robust and reliable components.

CHAPTER 6: THE WHEEL:
AN INFLATABLE SPACE STATION

1. Edward Everett Hale, *The Brick Moon and Other Stories* (Boston: Little, Brown, 1899).

2. Konstantin Tsiolkovsky, *Beyond the Planet Earth*, trans. Kenneth Syers (New York: Pergamon, 1960).

3. When Oberth was later working with the US after WWII, he suggested that an orbiting mirror as large as 300 *miles* in diameter could be useful for peaceful purposes. He was what we might refer to today as a "big thinker."

4. "The German Space Mirror," *Life*, July 23, 1945, 78–80.

5. Hermann Noordung (Potocnik), *The Problem of Space Travel* (Washington, DC: NASA, 1995).

6. Ley had already achieved broad notoriety for, among other things, a book called *The Conquest of Space*, published in 1949, which covered much of the same material in some detail.

7. Willy Ley, "A Station in Space," *Collier's*, March 22, 1952, 30–31.

8. Though, it's worth noting that this was almost exactly the same time that von Braun's *The Mars Project* was being published, with its 950-launch assembly schedule, so it may have simply been more indicative of the directions of his passions.

9. Wernher von Braun, "Crossing the Last Frontier," *Collier's*, March 22, 1952, 24–29, 72–74.

10. Ibid.

11. Ibid.

12. Ibid.

13. A slightly later 1955 Disney production, complete with extensive animation and live-action simulations of a lunar journey, makes mention of Johnston Atoll. The film contains an animated zoom from the air on the islands, referring to them as "a small atoll of coral islands in the Pacific, where man is dedicated to just one cause … the conquest of space." Ward Kimball and William Bosche, "Man in Space," *Disneyland*, season 1, episode 20, directed by Ward Kimball, aired March 9, 1955.

14. Michael Belfiore, "Bigelow Aerospace's Inflatable Space Station," *Bloomberg*, May 12, 2016, http://www.bloomberg.com/news/articles/2016-05-12/bigelow-aerospace-s-inflatable-space-station (accessed August 31, 2016).

15. For more information on the BEAM project, see NASA's BEAM website: http://www.nasa.gov/mission_pages/station/research/experiments/1804.html.

CHAPTER 7: VENUSIAN EMPIRE: NASA'S MARS/VENUS FLYBY ADVENTURE

1. David S. F. Portree, "Humans to Mars: 50 Years of Mission Planning, 1950–2000," *NASA Monographs in Aerospace History* no. 21, NASA SP-2001-4521 (Washington, DC: NASA, February 2001), pp. 11–12.

2. Ibid., p 22.

3. The budget in 1962 was thirteen times that in NASA's founding year, 1958, and five and a half times that of 1959, the agency's first full year of operations—about 1.18 percent of the federal budget. The peak was in 1966, at 4.41 percent. Lunar and Planetary Institute, http://www.lpi.usra.edu/exploration/multimedia/NASABudgetHistory.pdf (accessed September 10, 2016).

4. Franklin P. Dixon, "The EMPIRE Dual Planet Flyby Mission," Aeronutronic Division, Philco Corporation; paper presented at the Engineering Problems of Manned Interplanetary Exploration conference, Palo Alto, CA, September 30–October 1, 1963. Many people called NASA "the NASA" in its early history, because its predecessor, the National Advisory Committee on Aeronautics (which was abbreviated NACA) had gone by the name "the NACA." Most of "the NACA" was absorbed into NASA when it was formed in 1958.

5. Earlier testing involved a smaller reactor called KIWI, and then it began with NERVA in 1964.

6. W. H. Robbins and H. B. Finger, "An Historical Perspective of the NERVA Nuclear Rocket Engine Technology Program," NASA Contractor Report 187154/AIAA-91-3451, NASA, July 1991.

7. "Nova," *Encyclopedia Astronautica*, http://www.astronautix.com/n/nova.html (accessed September 10, 2016). There were many planned variations on the Nova, the largest of which would be able to loft almost half a million pounds.

8. David S. F. Portree, "EMPIRE Building: Ford Aeronutronic's Mars/Venus Piloted Flyby Study (1962)," *WIRED*, January 23, 2013, http://www.wired.com/2013/01/ford-aeronutronic-empire-1962/ (accessed September 1, 2016).

9. This would have also acted to evenly distribute the temperature extremes encountered in spaceflight across the hull, and was an approach (sans the gravity-inducing habitation units) used during the Apollo flights to the moon. For that two-day voyage, the entire spacecraft was put into a slow spin, which was called "barbecue mode," to even out heating and cooling

from the direct rays of the sun. In a long-duration flight, especially toward Venus, this would have been of great import.

10. Portree, "EMPIRE Building."

11. Dixon, "The EMPIRE Dual Planet Flyby Mission."

12. "Project Apollo: A Retrospective Analysis," http://history.nasa.gov/Apollomon/Apollo.html#note36 (accessed August 12, 2016).

13. Dixon, "The EMPIRE Dual Planet Flyby Mission."

14. Frederick I. Ordway III, Mitchell R. Sharpe, and Ronald C. Wakeford, "EMPIRE: Early Manned Planetary-Interplanetary Roundtrip Expeditions Part I: Aeronutronic and General Dynamics Studies," *Journal of the British Interplanetary Society* (May 1993): 179–90.

15. Quote attributed to Frank Borman, Apollo 204 Review Board Testimony, 1967.

16. Mariner 2 was a modified version of the early Ranger probes sent to the moon, and carried only basic instrumentation. It was essentially an engineering test flight with some scientific data returned.

17. Harry O. Ruppe, "Manned Planetary Reconnaissance Mission Study: Venus/Mars Flyby," NASA, February 5, 1965.

18. Ibid; Annie Platoff, "Eyes on the Red Planet: Human Mars Mission Planning, 1952–1970," NASA.

19. Platoff, "Eyes on the Red Planet."

20. W. B. Thompson and J. E. Volonte, "Experiment Payloads for Manned Encounter Missions to Mars and Venus," Bellcom, 1967.

21. Ruppe, "Manned Planetary Reconnaissance"; "Project Apollo: A Retrospective Analysis," NASA, http://history.nasa.gov/Apollomon/Apollo.html (accessed September 10, 2016).

22. Edward A. Willis, Jr., Comparison of Trajectory Profiles and Nuclear-Propulsion-Module Arrangements for Manned Mars and Mars-Venus Missions (Cleveland, OH: Lewis Research Center, 1971).

23. Ibid.

CHAPTER 8: BLUE GEMINI: WEAPONIZING ORBIT

1. "X-20 Dyna-Soar Space Vehicle," Boeing, http://www.boeing.com/history/products/x-20-dyna-soar.page (accessed September 14, 2016); Roy

F. Houchin, *US Hypersonic Research and Development: The Rise and Fall of Dyna-Soar* (New York: Routledge, 2006).

    2. Dwayne A. Day, "The Blue Gemini Blues," *Space Review*, March 20, 2006, http://www.thespacereview.com/article/582/1 (accessed September 14, 2016); Clarence J. Geiger, "History of the X-20A Dyna-Soar, Volume I," AFSC, October 1963, http://www.dtic.mil/dtic/tr/fulltext/u2/a951933.pdf (accessed September 14, 2016).

    3. Neil Sheehan, *A Fiery Peace in a Cold War: Bernard Schriever and the Ultimate Weapon* (New York: Vintage Books, 2010), p. 3.

    4. Ibid., p. 56.

    5. For more information on the AMU program: Daniel D. McKee, *The US Air Force in Space*, ed. Eldon W. Downs (New York: Frederick A. Praeger, 1966), pp. 6–15; "Gemini Astronaut Maneuvering Unit Ready Reference Handbook," http://web.archive.org/web/20040218143025/http://www.thebest.net/jduncan/geminiamu/geminiamu.htm (accessed September 14, 2016).

    6. "Blue Gemini," http://www.hq.nasa.gov/office/pao/History/SP-4203/ch6-2.htm#source7 (accessed September 1, 2016).

CHAPTER 9: FLIRTING WITH DEATH: THE TERRIFYING FLIGHT OF GEMINI 8

    1. David Scott and Alexei Leonov, *Two Sides of the Moon: Our Story of the Cold War Space Race* (New York: Thomas Dunne Books, 2004), p. 165.

    2. "Gemini VIII Voice Communications (Air-to-Ground, Ground-to-Air and On-Board Transcription)," NASA, p. 71, http://www.jsc.nasa.gov/history/mission_trans/GT08_TEC.PDF (accessed September 15, 2016).

    3. Scott and Leonov, *Two Sides of the Moon*, p. 166.

    4. "Gemini VIII Voice Communications," p. 74.

    5. Scott and Leonov, *Two Sides of the Moon*, p. 166.

    6. "Gemini VIII Voice Communications," p. 75.

    7. Ibid.

    8. Ibid.

    9. Ibid.

    10. During the Gemini program, the standard maneuvering thrusters were called the Orbital Attitude and Maneuvering System. The reentry

thrusters, intended for use only in returning from the mission, were abbreviated as the reentry control system, or RCS. By the time the Apollo missions were underway, the normal maneuvering thrusters were called the reaction control system or RCS.

    11. "Gemini VIII Voice Communications," p. 78.

CHAPTER 10: MANNED ORBITING LABORATORY: HOW TO DESIGN, TEST, AND NEVER FLY A SPACE PROGRAM

    1. A compendium of the MOL program's declassified files can be seen on the NRO's online database at: http://www.nro.gov/history/csnr/programs/docs/MOL_Compendium_August_2015.pdf.

    2. Ibid.

    3. Department of Defense, "Air Force to Develop Manned Orbiting Laboratory," NRO Archives, December 10, 1963.

    4. James D. Outzen, ed., *The Dorian Files Revealed: A Compendium of the NRO' s Manned Orbiting Laboratory Documents* (Chantilly, VA: National Reconnaissance Office, 2015), p. 76, http://www.nro.gov/history/csnr/programs/docs/MOL_Compendium_August_2015.pdf (accessed September 2, 2016). As Arthur Goldberg was the US ambassador to the United Nations, it has been suggested by historians that when LBJ said "United States" he had meant to say "United Nations."

    5. Ibid., p. 17.

    6. Dean Rusk, in a letter to Robert McNamara, "Subject: 7/6/65 Space Council Discussion of the MOL Project," NRO Archives, 1965.

    7. Outzen, *Dorian Files Revealed*, p. xiv.

    8. These satellite program names are frequently seen as CORONA, GAMBIT, HEXAGON, etc. These names are not acronyms, they are code-names, and as such should not be in all capitals, according to US Air Force sources.

    9. Dwayne A. Day, "All Alone in the Night," *Space Review*, June 23, 2014, http://www.thespacereview.com/article/2539/1 (accessed September 15, 2016).

    10. Letter from Robert McNamara to Vice President Johnson, NRO Archives, August 9, 1963.

11. Dr. Harold Brown, testimony before the Senate Committee on Aeronautical and Space Sciences, 89th Cong., 2nd Session, 1965 NASA Authorization (March 9, 1964), pt. 2, p. 461.

12. Outzen, *Dorian Files Revealed*, p. 53.

13. "The Gemini Program (1962–1965)," *NASA*, http://nssdc.gsfc.nasa.gov/planetary/gemini.html (accessed September 6, 2016).

14. Department of Defense, Proceedings of Air Force MOL Policy Committee Meeting, NRO Archives, April 29, 1966.

15. Outzen, *Dorian Files Revealed*, p. 101.

16. Schriever to SAF, memorandum, "Subject: Manned Mil Missions in Space," August 31, 1966.

17. Outzen, *Dorian Files Revealed*, p. 148.

18. Ibid., p. 148–49.

19. "Nixon, Defending Policy, Hits 'New Isolationists'; Pledges a World Role; He Chides Critics," *New York Times* (June 5, 1969): 1.

20. Outzen, *Dorian Files Revealed*, p. 164.

21. National Reconnaissance Office, http://www.nro.gov/foia/declass/HEXAGON.html (accessed September 15, 206).

22. David Winfrey, "The Last Spacemen: MOL and What Might Have Been," *Space Review*, November 16, 2015, http://www.thespacereview.com/article/2866/1 (accessed September 15, 2016).

23. "Advanced MOL Planning," 1965, http://www.nro.gov/foia/declass/mol/794.pdf (accessed September 15, 2016).

CHAPTER 11: APOLLO 11: DANGER ON THE MOON

1. This is true across the US space program in general; one exception was the selection of North American Aviation (later North American Rockwell) to build the command and service modules—there was reportedly a bit of last-minute maneuvering within NASA's upper brass to sway the final decision away from Martin Aviation (which scored higher in their initial evaluation) and toward NAA, which when later revealed caused a bit of controversy. NAA had rated higher in "technical qualifications," but Martin had rated higher in "technical approach." NASA/MSC, "Source Evaluation Board Report," pp. 10–14.

2. Michael Collins, *Carrying the Fire: An Astronaut's Journeys* (New York: Cooper Square, 2001), p. 314.

3. *Broadcasting Magazine*, July 28, 1969, republished in "A Remote that Broke All the Records," *Broadcasting and Cable*, July 17, 2009, http://www.broadcastingcable.com/news/news-articles/remote-broke-all-records/110227?rssid=20065 (accessed September 6, 2016).

4. For all flight dialogue and quotes: Eric M. Jones, ed., "Apollo 11 Lunar Surface Journal," NASA, https://www.hq.nasa.gov/alsj/frame.html (accessed September 15, 2016).

5. Gene Kranz, in an interview with the author, Johnson Space Center, 2005.

6. Buzz Aldrin, in an interview with the author, 2005.

7. Steve Bales, in an interview with the author, 2005.

8. Aldrin, interview with the author.

9. Ibid.

10. Charlie Duke, in an interview with the author, 2005.

11. Bales, interview with the author.

12. Aldrin, interview with the author.

13. Kranz, interview with the author.

14. Aldrin, interview with the author.

15. Kranz, interview with the author.

16. Dick Dunne, in an interview with the author, 2005.

17. Aldrin, interview with the author.

18. Kranz, interview with the author.

19. Ibid.

20. Smoking was banned in federal buildings, including Mission Control, in 1997.

21. Kranz, interview with the author.

CHAPTER 12: THE FIRST SPACE SHUTTLE: PROJECT DYNA-SOAR

1. While specific US Air Force projects to build its own crewed spacecraft technically ended with the cancellation of the Manned Orbiting Laboratory in 1969, the air force's goals had a strong hand in the design of the space shuttle in the 1970s. During the early years of that program, the air force flew classified payloads—and some air force astronauts—aboard the shuttle. The air force returned to using expendable rockets after the loss of

Challenger, and eventually developed the completely robotic (and reportedly very successful) X-37 mini-spaceplane, which launches on an Atlas V.

2. "History of the X-20 Dyna-Soar," Air Force Systems Command, Historical Publications Series 63-50-I, 1963.

3. The "Karman Line," the generally accepted boundary of space, is about sixty-two miles. Both NASA and the US Air Force, however, defined "space" as fifty miles altitude at the time.

4. Matthew A. Bentley, *Spaceplanes: From Airport to Spaceport* (New York: Springer, 2009), p. 117.

5. The X-20's mission changed a few times over its short life, and one of the largest shifts was considering whether its role was primarily as a suborbital long-range glider or an orbital craft. See "History of the X-20 Dyna-Soar," Air Force Systems Command, Historical Publications Series 63-50-I, 1963.

6. Ibid.

7. NASA, *Facing the Heat Barrier: A History of Hypersonics*, p. 143, http://history.nasa.gov/sp4232-part2.pdf (accessed August 23, 2016).

CHAPTER 13: BEYOND THE EDGE OF SPACE: THE X-15B

1. "The First Suborbital Spaceplane," video recording of statements by Joe Engle to Alvin Remmers, MoonandBack Inc., 1989.

2. "NASA Armstrong Fact Sheet: X-15 Hypersonic Research Program," *NASA*, February 28, 2014, https://www.nasa.gov/centers/armstrong/news/FactSheets/FS-052-DFRC.html (accessed August 23, 2016).

3. The actual measurement of the most-used version of the X-15 was forty-nine feet, two inches.

4. Ibid.

5. NASA's and the US Air Force's accepted definition of space was fifty miles altitude in the 1960s.

6. Ibid.

7. Michelle Evans, *The X-15 Rocket Plane: Flying the First Wings into Space* (Lincoln, NE: University of Nebraska Press, 2013).

8. Ibid.

9. Ibid.

10. Peter W. Merlin, "Michael Adams: Remembering a Fallen Hero," *NASA:*

People and Places, July 30, 2004, http://www.nasa.gov/centers/dryden/news/ X-Press/stories/2004/073004/ppl_adams.html (August 23, 2016).

11. Michelle Evans, in a conversation with the author, 2016.

12. L. Parker Temple, "X-15B: Pursuit of Early Orbital Human Spaceflight," *Air Power History* 55, no. 1 (March 2008): 29–41.

13. Joe Engle, in an interview with the author, 2014.

14. While not known outside the Soviet Union at the time, this was similar to the technique that the Soviets were using for their first spacecraft, the Vostok: when the capsule was at a low altitude over Soviet territory, the cosmonaut ejected from the capsule, which would parachute down to terra firma. The cosmonaut would be spared the rougher touchdown of the heavy capsule, drifting to the ground nearby.

15. Temple, "X-15B," pp. 29–41.

16. Engle, interview with the author.

CHAPTER 14: THE SAD, STRANGE TALE OF SOYUZ 1

1. Pat Norris, *Spies in the Sky: Surveillance Satellites in War and Peace* (New York: Springer), p. 7.

2. While the original designation for the flight was simply *Vostok*, *Vostok* 1 is used here for clarity.

3. "R-7 Ballistic Missile," *Russian Space Web*, http://www.russian spaceweb.com/r7.html (accessed September 19, 2016).

4. Ben Evans, "'This Devil Ship': The Tragic Tale of Soyuz 1," *America Space*, April 24, 2012, http://www.americaspace.com/?p=17776 (accessed September 19, 2016).

5. Andrew Chaikin, *A Man on the Moon: The Voyages of the Apollo Astronauts* (New York: Penguin, 2007).

6. "Soyuz 7K-OK," *Encyclopedia Astronautica*, http://www.astronautix. com/s/soyuz7k-ok.html (accessed September 19, 2016).

7. For this information and the next few paragraphs: "The Mission of Soyuz-1," NASA, http://spaceflight.nasa.gov/outreach/SignificantIncidents/ assets/the-mission-of-soyuz-1.pdf (accessed September 19, 2016).

8. "Soyuz 1," *Space Safety Magazine*, http://www.spacesafetymagazine. com/space-disasters/soyuz-1/ (accessed September 19, 2016).

9. Associated Press, "Yuri Gagarin Killed as Test Plane Falls," *New York Times*, March 28, 1968, http://www.nytimes.com/learning/general/onthisday/bday/0309.html.

10. Nikolai Tsymbal, ed., *First Man in Space* (Moscow: Progress Publishers Moscow, 1984).

11. Ibid.

CHAPTER 15: THE TURTLENAUTS

1. National Intelligence Estimate number 11-1-65 (supersedes NIE 11-1-62), Soviet Space Program, submitted by the director of Central Intelligence, concurred in by the United States Intelligence Board, January 27, 1965.

2. Ibid.

3. Ibid.

4. Sven Grahn, "Jodrell Bank's Role in Early Space Tracking Activities—Part 2," http://www.svengrahn.pp.se/trackind/jodrell/jodrole2.htm (accessed September 8, 2016).

5. Frank Borman and Robert Sterling, *Countdown: An Autobiography* (New York: Silver Arrow Books, 1988).

6. Dwayne Day, "Chasing Shadows: Apollo 8 and the CIA," *Space Review* (April 11, 2016).

CHAPTER 16: FALLING TO EARTH: THE DANGEROUS SCIENCE OF REENTRY

1. The crew of Soyuz 11 also died during reentry, but this was due to an air valve opening prematurely and venting their capsule. The resulting near-vacuum asphyxiated the crew of three in 1971.

2. The shuttle was configured to carry seven astronauts, but carried eight on one mission, STS-61-A. In an emergency, contingency plans cited carrying up to ten people.

3. Stephen Smith, "Suit Tossed During Spacewalk," CBS News, February 3, 2006, http://www.cbsnews.com/news/suit-tossed-during-spacewalk/ (accessed August 25, 2016).

4. Ray Bradbury, *The Illustrated Man* (New York: Doubleday, 1951).

CHAPTER 17: FUNERAL FOR A VIKING: THE END OF VIKING 1

1. Rachel Tillman (founder and executive director of the Viking Mars Missions Education and Preservation Project), in an interview with the author, 2015.

2. William R. Corliss, *The Viking Mission to Mars* (Washington, DC: NASA, 1974).

3. "Interview with Norman H. Horowitz," interview by Rachel Prud'Homme, California Institute of Technology, 1987.

4. Ibid.

5. Edward Clinton Ezell and Linda Neuman Ezell, *On Mars: Exploration of the Red Planet 1958–1978* (Washington, DC: NASA, 1984). The intimation is that while they were able to see things the size of the Rose Bowl, and much smaller, they were still a long way from seeing eighteen-inch craters.

CHAPTER 18: SAVING SKYLAB: COWBOYS IN SPACE

1. Elizabeth Howell, "Eugene Cernan: Last Man on the Moon," *Space.com*, April 23, 2013, http://www.space.com/20790-eugene-cernan-astronaut -biography.html (accessed September 21, 2016).

2. Four flights for the Skylab program, and one for Apollo-Soyuz. The only additional Saturn V launched Skylab, and the rest of the missions were launched with the Saturn IB.

3. "Soyuz 11: Triumph and Tragedy," NASA, http://www.hq.nasa.gov/ pao/History/SP-4209/ch5-6.htm (accessed September 8, 2016).

4. Many differing measurements of Skylab are listed by NASA and other resources, and depend on which configuration is being considered. The values given here are of the basic "can" that comprised the space station without solar panels extended.

5. The flight backup is at the Smithsonian's National Air and Space Museum in Washington, DC, and two training mock-ups are at the Johnson Space Center's Space Center Houston and the US Space and Rocket Center in Huntsville, Alabama.

6. W. David Compton and Charles D. Benson, *Living and Working in Space: A History of Skylab* (Washington, DC: NASA, 1983).

7. David Hitt, Owen Garriott, and Joe Kerwin, *Homesteading Space: The Skylab Story* (Lincoln, NE: University of Nebraska Press, 2008), p. 141.

8. "NASA Johnson Space Center Oral History Project: Dr. Joseph P. Kerwin," interview by Kevin M. Rusnak, NASA, May 12, 2000, http://www.jsc.nasa.gov/history/oral_histories/KerwinJP/KerwinJP_5-12-00.htm (accessed September 20, 2016).

9. Ibid.

10. Ben Evans, *At Home in Space: The Late Seventies into the Eighties* (New York: Springer, 2012), p. 170.

11. Ibid., p. 78.

12. Ibid., p. 171.

13. Ibid., pp. 172–73.

14. Ibid., p. 174.

15. "NASA Johnson Space Center Oral History Project: Kerwin."

16. Hitt, Garriott, and Kerwin, *Homesteading Space*, p. 385.

CHAPTER 19: NEAR MISSES: DANGER STALKS THE SPACE SHUTTLE

1. In another twist of irony, the Space Launch System, the mega-booster that NASA is working on today, uses shuttle SSMEs for its rocket, and they are not returned to Earth for reuse. They are thrown away with the rest of the rocket. It should be noted, however, that since they are not required to support multiple missions, alternative fabrication techniques are being used in their construction that allows new power plants to be built for less cost.

2. Mark Wade, "Space Shuttle," *Encyclopedia Astronautica*, http://www.astronautix.com/s/spaceshuttle.html (accessed September 20, 2016).

3. To be fair, this includes downtime suffered after the *Challenger* and *Columbia* accidents, so this figure is somewhat misleading. But at its best, the shuttle never came close to its promised flight frequency.

4. Carl Bialik, "As Shuttle Sails Through Space, Costs Are Tough to Pin Down," *Wall Street Journal*, July 9, 2011, http://www.wsj.com/articles/SB10001424052702303544604576433830373220742.

5. Richard P. Feynman, "Appendix F–Personal Observations on the Reliability of the Shuttle," Report of the Presidential Commission on the Space Shuttle *Challenger* Accident, NASA, June 1986.

6. William Harwood, "Legendary Commander Tells Story of Shuttle's Close Call," CBS News Space Place, March 27, 2009.

7. Feynman, "Appendix F–Personal Observations."

8. Mark Garcia, ed., "Space Debris and Human Spacecraft," NASA, September 26, 2013, http://www.nasa.gov/mission_pages/station/news/orbital_debris.html (accessed September 20, 2016).

9. Robert Lee Hotz, "Harmless Debris on Earth Is Devastating in Orbit," *Wall Street Journal*, February 27, 2009, http://www.wsj.com/articles/SB123568403874486701.

CHAPTER 20: SHOWDOWN IN SPACE:
FIREARMS ON THE MOON

1. Directorate of R&D, Future Weapons Office, US Army Weapons Command, "The Meanderings of a Weapon Oriented Mind When Applied in a Vacuum Such as on the Moon" (Rock Island, Illinois: Headquarters, US Army Weapons Command, July 1965).

2. Ibid.

3. Ibid.

4. For the literally minded, there are also high crater rims and mountains on the moon that would likely intercept a projectile fired at ground level before it completed an orbit. But it's possible, theoretically, that if fired from a proper elevation at a proper angle, that this would not occur.

5. Actual diameter of the rounds seems to have varied from .49–.51 caliber, or 12–13mm, depending on the iteration of the weapon.

6. The records of what exact weapon was incorporated into Salyut 3 are murky. The general consensus seems to indicate that the weapon was a Rikhter 23, a modified 23mm aerial cannon, while other records discuss the Nudelman-Rikhter NR 30, a 30mm gun.

7. Michael Peck, "Revealed: The Soviet Union's Space Cannon," *National Interest*, October 14, 2015, http://nationalinterest.org/feature/revealed-the-soviet-unions-space-cannon-14068 (accessed September 21, 2016). Accounts vary as to the caliber of the weapon.

8. Directorate, "Meanderings of a Weapon Oriented Mind."

CHAPTER 21: *BURAN*: THE SOVIET UNION'S ONE-FLIGHT WONDER

1. Anatoly Zak, *Russian Space Web*, http://www.russianspaceweb.com (accessed July 2016).

2. SpaceX's Dragon capsule can return small amounts of cargo to Earth, but did not begin flying until the twenty-first century.

3. The existence of these weapon systems is arguable, but they are reported to have been designed and, in some cases, tested—at least on the ground.

4. Robyn Dixon, "Eight Workers Feared Dead in Russian Cosmodrome," *Los Angeles Times*, May 13, 2002, http://articles.latimes.com/2002/may/13/world/fg-russia13.

5. Office of the Secretary of Defense, "Soviet Acquisition of Militarily Significant Western Technology: An Update," publication AD-A160 564, Washington, DC, 1985. Cited in part from original publication and an article by Dwayne Day, "Theft, the Sincerest Form of Flattery," *Space Review*, April 16, 2012.

6. Ibid.

7. Ibid.

8. Robert Windrem, "How the Soviets Stole a Space Shuttle," *NBC News Space Report*, November 4, 1997.

9. Ibid.

10. Ibid.

CHAPTER 22: MAJOR MATT MASON: A MAN FOR THE NEW SPACE AGE

1. NASA did make institutional desegregation a priority by the mid-1960s, however. "Chapter IV: The Marshall Reconstruction," NASA, http://history.msfc.nasa.gov/book/chptfour.pdf (accessed September 21, 2016).

2. Henry Samuel, "European Space Agency Unveils 'Lunar Village' Plans as Stepping Stone to Mars," *Telegraph*, January 15, 2016, http://www.telegraph.co.uk/news/worldnews/europe/france/12102673/European-Space-Agency-unveils-lunar-village-plans-as-stepping-stone-to-Mars.html.

3. "Falcon 9," *SpaceX*, http://www.spacex.com/falcon9 (accessed Sep-

tember 12, 2016); "Falcon Heavy," *SpaceX*, http://www.spacex.com/falcon-heavy (accessed September 12, 2016).

4. Nicky Woolf and Amanda Holpuch, "One Pilot Dead as Virgin Galactic's SpaceShipTwo Rocket Plane Crashes," *Guardian*, November 1, 2014, https://www.theguardian.com/science/2014/oct/31/spaceshiptwo-richard-branson-virgin-crash-mojave.

5. Adam Chitwood, "*Major Matt Mason* Moves Forward with Tom Hanks and Robert Zemeckis," *Collider*, October 15, 2012, http://collider.com/tom-hanks-robert-zemeckis-major-matt-mason-moves-forward/ (accessed August 30, 2016).

INDEX

Page numbers in **bold** indicate photos and illustrations.

OAMS. *See* Orbital Attitude and Maneuvering System

Oberth, Hermann, 10–11, 12, 23, 83, 84, 109, 309n3

Okarian (Humvee), 8, 52, 53

Orbital Attitude and Maneuvering System (OAMS), 124, 125, 312n10

Orion, Project
        ending of, 66

Orion, Project (nuclear pulse ship), 8, 57–67, **60**, **63**, 93, 95, 160, 307n16
        diagram of propulsion system, **62**
        fallout dangers, 63–64, 306n9
        Orion Mk. II sized to fit on Saturn V, 64–65, 66
        specifications for, 61–62, 306n5

Outer Space Treaty of 1967, 266

Overmyer, Robert, 143

Pal, George, 86

parabolic mirror, orbiting, 83–84, 309n3

parachute failures, 213
        on Hexagon satellite, 143
        on the Soyuz 1, 197, 198, 213

Paracone, 215, **217**

Passage to Mars (documentary), 53

Paths to Space Travel, The (Oberth), 83

payload bays, 164, 180, 252, 281, 283, 286

PerkinElmer (company), 143

Phillips, Samuel, 79, 308n8

Planetary Resources (company), 295

plutonium, use of in spacecraft power, 307n15

Potocnik, Herman (aka Hermann Noordung), 84, 85

Pravda (newspaper), 198, 202

pressure suits, **33**, 52, 53, 172, 173, 189, 216–17, 243, 268

Problem of Space Travel, The (Noordung), 84

Project Mars: A Technical Tale (von Braun), 43

Projects. *See inverted names, i.e.,* Gemini, Project; Horizon, Project, *etc.*

Proton rocket, 204, 205, 206, 287

Pueblo, North Korea's capture of, 142

R-7 ICBM repurposed for Sputnik, 186, 187

radiation in space, 20, 43, 50, 67, 77, 97, 98, 181, 211, 306n9

ramjet engines, 12, 174

Ranger probes modified for Mariner 2, 311n16

RCC. *See* reinforced carbon composite

RCS. *See* Reentry Control System

Reaction Motors, 171

Reagan, Ronald, 275, 289

Redstone rockets, 25, 29, 97–98

reentry
        alignment, 211
        for Apollo missions, 208
        dangerous science of, 211–18
        problems with for Zond V, 208
        rockets on Gemini 8, 128

Reentry Control System (RCS) on Gemini 8, 126–28
        renamed Reaction Control System (RCS) in Apollo missions, 312n10

refracting telescopes, 37, 304n3

reinforced carbon composite (RCC), 255–56

reliability, importance of, 76, 307n4

René 41 alloy, 162, 180, 215

Rescue Gemini, 214, 318n2

rescue plans in case of emergencies, 213–18
        illustration of planned rescue mission for Skylab, **248**

retrorockets, 213

Revell company, 131

Rikhter cannon, 276, 321n6

RL Mark VI Space Diving Ensemble, 216–18

RoBo (Rocket Bomber) program, 159

robotics, **132**, 136, 221, 229, 284, 295, 306n11, 315n1
        mining, 295
        superiority of for orbital espionage, 140–41, 187, 293
        use of to scan for damage, 256–57

rocket fuels
        atomic rockets in Project Orion, 57–67

See also Dyna-Soar, Project; X-15
suborbital rocket plane

Yeager, Chuck, 157, 174–75

Zond project, 205, 206–210
compared to Apollo system, 206
Zond 4, 207

Zond 5, 207–208
capsule design for, **207**
carrying plants and animals,
208–209
Zond 6 and 8, 209
Zuckert, Eugene, 133–34